SpringerBriefs in Statistics

For further volumes:
http://www.springer.com/series/8921

Edgar Santos-Fernández

Multivariate Statistical Quality Control Using R

Springer

Edgar Santos-Fernández
Marketing and Communication Group Head
Empresa de Telecomunicaciones de Cuba S.A. (ETECSA). Villa Clara.
D #17 between Carr. Camajuani and 1st. Santa Catalina.
Santa Clara. 50300, Cuba

ISSN 2191-544X ISSN 2191-5458 (electronic)
ISBN 978-1-4614-5452-6 ISBN 978-1-4614-5453-3 (eBook)
DOI 10.1007/978-1-4614-5453-3
Springer New York Heidelberg Dordrecht London

Library of Congress Control Number: 2012947029

Printed on acid-free paper

Springer is part of Springer Science+Business Media (www.springer.com)

Preface

Nowadays, the intensive use of an automatic data acquisition systems and the use of on-line computers for process monitoring have led to an increased occurrence of industrial processes with two or more correlated quality characteristics, in which the statistical process control and the capability analysis should be performed using multivariate methodologies.

Unfortunately, despite the availability of increased computing capabilities, in the Multivariate Statistical Quality Control (MSQC) framework the software solutions are limited or restricted in their level of success and ease of use for dealing with the problems of industry or promoting academic instruction.

The aim of this book is to present the most important MSQC techniques developed in R language, across the most important theoretical aspects (without pretending to be a book in statistical theory) of the use of the software and the solution of problems. The choice of R is motivated by the fact that the R language has become the "lingua franca" of the data analysis and it is an easy-to-use, open source, free, multiplatform, and very flexible software. Further, R has a mounting community of users; it has been growing up in solutions for corporations and the acceptance in the academia.

This is a succinct, comprehensible and accessible text that provides the core of the MSQC tools across illustrative examples done by hand and using computer software presenting the code snippets. The following word cloud shows the main topics approached in this book in proportion to the font size.

The first chapter provides a very short introduction to R language, statistical procedures, and the main aspects concerning Statistical Quality Control (SQC).

Chapters 2 and 3 constitute the kernel of this book in which the design and interpretation of multivariate control chart and the computation of multivariate process capability indices are covered. Chapter 5 approaches the tools for assessing multivariate normality and independence, and Chap. 6 contains two study cases integrating the knowledge acquired in previous sections. This text could be read in the order desired by the reader.

Ideal to postgraduate courses in SQC, Quality Engineering, Industrial Statistics, and Industrial Engineering it could nonetheless be used for advanced undergraduate

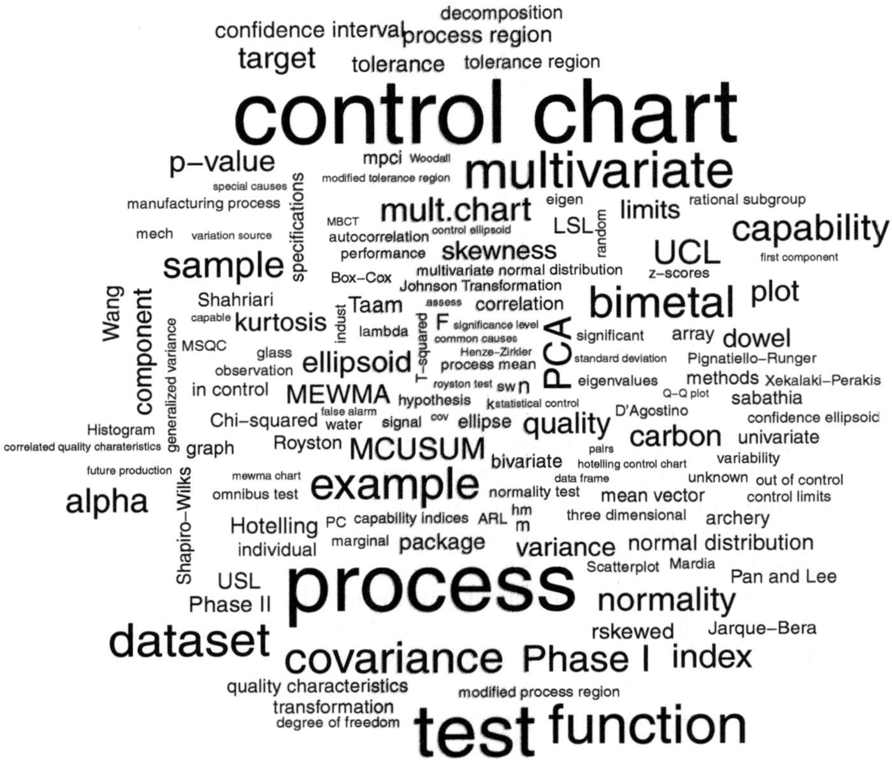

Fig. Word cloud of the main subjects

students. It includes the MSQC R package, available at http://www.cran.r-project.org/package=MSQC from CRAN (the Comprehensive R Archive Network) and holds the eleven dataset used.

The examples code and the solutions to all exercises are available at the author web page (https://sites.google.com/site/edgarsantosfernandez/). This site can be also consulted for additional information and for the list of errata.The reports of suggestions, errors or omissions are most welcome at: edgar.santos@etecsa.cu.

The book assumes the reader has an elemental background in matrix algebra, statistics, and practically no computer skills in R.

It provides statisticians, scientists, engineers, practitioners, and students a modern and practical overview about the most accepted techniques on MSQC of the last years across the examples and exercises.

In other words, it supplies the knowledge and the computational tools necessary for solving the main problems presented in this field and for practically nothing.

Santa Clara, Cuba Edgar Santos-Fernández

Acknowledgments

I am immensely special thanks to the R core team and contributors for the creation and development of the R language and environment; to the Springer Team Marc Strauss, Hannah Bracken and Susan Westendorf who made this project possible; and to my friend and mentor Michele Scagliarini for his collaboration and help, and for being a true coauthor of the mpci function, without which this text could have never been possible.

I would like to acknowledge the contribution of Professors William H. Woodall and Ramanathan Gnanadesikan and also Matias Salibian-Barrera, Pat Farrell, Scott Ulman, Surajit Pal, Uwe Ligges, Kurt Hornik, Achim Zeileis, and Michel Marrero-González.

I also thank Ricardo Reyes Perera and Maritza García Pallas for the revisions of this book. I would also like to give special thanks to the Universidad Central de Las Villas and to the Empresa de Telecomunicaciones de Cuba S.A. (ETECSA) for the support.

Heartfelt thanks to Carmen Fernández Ferrer to whom this book is dedicated for the inspiration and to Caroline, Laura and Alex and to my other siblings Isis and Alejandro for the enthusiasm, to Silvio for the encouragement, to Jessica for the joy she brings to my life, and to my father Eugenio Santos Miyares (Cuqui) (1950–2011) and Georgina Ferrer Riera (1924–2000) whom are always in our memory.

In short, to all those who contributed to this project and to my family for the patience and the support during the writing of this book.

Contents

Chapter 1
A Small Introduction

1.1 A Small Introduction

1.1.1 A Brief on R

R is high-level and open-source programming language focused mainly in statistical processing. It is based on the recognized S language and allows the integration with others as C, C++, Fortran, Java, Python, etc.

There are many characteristics which have placed it in the elite of the statistical computing software. It is an easy-to-use, flexible, and powerful software with an excellent performance regarding its competitors. Besides it is multiplatform, that is, runs over UNIX, Windows, and Mac OS. Moreover, and last but not least, it is absolutely free; contrasting with the high cost of similar proprietary software.

Another remarkable feature is that it constitutes one of the biggest knowledge and technology transfer to developing countries.

The R software per se consists barely in a few megabytes including the basic function, which is frequently updated. This philosophy allows a lightly main program kept by the user with only additional applications called packages.

These packages are available through the Comprehensive R Archive Network (CRAN).

Applications in R cover a wide range of disciplines such as Bioinformatics, Econometrics, Environmetrics, etc.

A remarkable feature of R is the huge community of users worldwide which have developed an extensive documentation and help sources including a mailing list with keen users.

A fact that upholds the above said is the exponential growth of literature about programming, graphics, etc. and the large amount of publications that refer applications or processing in R.

E. Santos-Fernández, *Multivariate Statistical Quality Control Using R*,
SpringerBriefs in Statistics 14, DOI 10.1007/978-1-4614-5453-3_1,

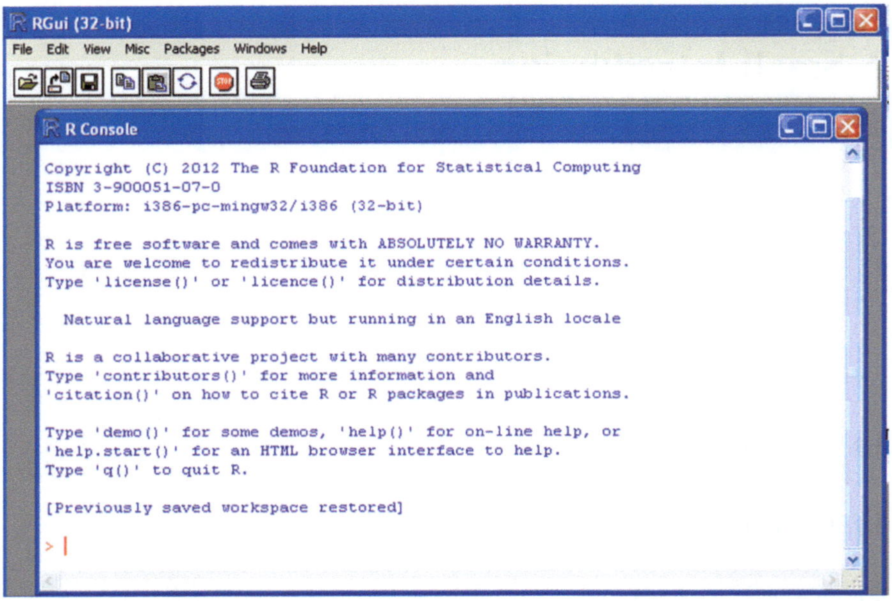

Fig. 1.1 R console in Windows

1.1.2 R Installation and Managing

The R installation is very simple. Just download the suitable version for your platform from a desired CRAN at http://cran.r-project.org/ and install it.

When R is opened, appears the R console with a message indicating the following information: the version, the platform, and the important statement that R comes without any warranty, the way to cite R and the packages in publications, etc. Besides that, in contributors() the R-core Team and contributors appear.

In this console the cursor is placed after the > symbol called *prompt* that indicates availability. On the other hand when + symbol appears, it means that the computation is not completed. Using the Up arrow key it is possible to invoke the last computation. One unique characteristic of the language is the assignment operator $<-$ instead of the symbol $=$ (Fig. 1.1).

All the content of this book has been produced in Windows, so any suggestion or report of inaccuracy is welcome.

All the examples in this book are contained in the MSQC package available at http://www.cran.r-project.org/package=MSQC and to install it just type:

> install.packages("MSQC")
selecting the desired CRAN.
Thus, to load it
> library("MSQC")

1.1.3 General Principles of Data Manipulation

The input data in R can be carried out in a simple way: using read.table: specifying the path

```
> data<-read.table("C:\\data.txt")
```
or
```
> data<-read.table(file.choose())
```
selecting the path where the file is located.

Besides files can be imported from another statistical software such as SPSS and MINITAB using the *foreign* package or from an Excel file with the *gdata* package.

Another useful input tool is the scan() function that makes it possible to read from the console. For instance:

```
> data <- scan()
1: 0.677
2: 0.852
...
11: 0.633
12: 0.637
13:
Read 12 items
```
This creates a vector named data with the 12 elements read.

In this book mainly three types of data structures are used: vectors, matrices, and arrays.

The vectors are the simplest structures in R. A vector can be composed by a unique element or by more than one, for instance vec <- 0 or vec <- 1:12.

A matrix is a two-dimensional set of data achieved using e.g.:

```
> data <- matrix(data, nrow = 6, ncol = 2)
```
Finally, an array allows a set of data with more than two dimensions.
```
> array(data, c(2,3,2))
```
produces a three-dimensional array

1.1.4 Datasets Used

Using the function data() it is possible to visualize all datasets included both in default datasets package and incorporated in other installed packages. The MSQC package includes the following datasets:

```
> data(package="MSQC")
```
dowel: Diameter and length of a manufacturing process of a dowel pin
carbon: Carbon fiber tubing

bimetal: A bimetallic strip used in a thermostat
industrial: A bivariate industrial process
water: A water quality test
mech: A mechanical process
glass: Glass manufacturing
rskewed: Right-skewed distribution
sabathia: A pitching log of C.C. Sabathia
archery: Target archery

1.1.5 The R Help

The R help is one of its strengths and can be exploited in different ways. When the function exists and the name is known, the most simple way is by using directly in the console the ? symbol followed by the function name. For example, after the installation of the MSQC package

> ? mult.chart

Then an html page opens showing elements of help such as usage, arguments, details, value, note, references, and examples. Occasionally the user ignores the exact number of the function then using help.search allows to search into the documentation database. For instance:

> help.search("capability indices") shows information about the mpci function.

Conversely in the main menu the option Help provides a lot of categories to evacuate doubts. The first cluster gives information about frequently asked question (FAQ) and pdf manuals.

The second refers to the previous help function introduced and the last one about the R-Project home page and CRAN.

R home page provides a lot of information in the section Documentation. Furthermore, the search on the web offers solutions to common problems.

Another important source about the operating are the examples incorporated at the end of the function documentation. They could be pasted directly to the console or using:

> example(MSQC)

Mathematical functions

Using ?S4groupGeneric; R returns the group of generic function with many categories e.g.: Arith, Compare, Logic, Math, Summary, etc.

Operators

The operator in R can be achieved as:
> ?Syntax
The most used ones in this book are:
arithmetics: +, -, *, /, ^,etc.
logical: <, >, <=, >=, ==, !=, etc.
and the component $

Table 1.1 Graphical function used in this book

Function	Description
plot	Scatterplot
qqnorm	Quantile–Quantile plot
barplot	Bar plot
pairs	Matrix of scatterplot
hist	Histogram

Table 1.2 Low level graphics

Function	Description
points	Add points by given coordinates
lines	Draws a line
rect	Draws a rectangle
arrows	Draws a arrow

Table 1.3 Some of the graphical parameters

Parameter	Description
lty	Line type
col	Colors
pch	Plotting symbol
mfrow, mfcol	Multiple graphs

1.1.6 Graphics in R

Another strength of the language is the high quality graphics produced. There are many Internet sites and books that cover the vast fields of graphics in R. The selection of the type of function to use depends on the nature of the data. The next table shows the main graphical function used in this book (Table 1.1).

Over an existing graph the forms presented in the following table can be added (Table 1.2):

By using help(par) parameters are obtained that can be used to customize the graphical representation (Table 1.3).

On the other hand, xlab and ylab allow the labeling of axes while xlim and ylim the coordinates ranges.

The graphics in R can be saved in many formats such as pdf, png, jpeg, bmp, postscript, etc. using for instance:

```
> postscript("foo.eps", width = 5.0, height = 4.0)
> plot(runif(20))
> dev.off()
```

or simply with a right-click on the graph and choosing copy or save.

Table 1.4 Built in probability distributions

Distribution	Density function	Distribution function	Quantile function	Random number generation
Beta	dbeta	pbeta	qbeta	rbeta
Binomial	dbinom	pbinom	qbinom	rbinom
Cauchy	dcauchy	pcauchy	qcauchy	rcauchy
Chi-squared	dchisq	pchisq	qchisq	rchisq
Exponential	dexp	pexp	qexp	rexp
F	df	pf	qf	rf
Gamma	dgamma	pgamma	qgamma	rgamma
Geometric	dgeom	pgeom	qgeom	rgeom
Hypergeometric	dhyper	phyper	qhyper	rhyper
Log-normal	dlnorm	plnorm	qlnorm	rlnorm
Multinomial	dmultinom	pmultinom	qmultinom	rmultinom
Negative binomial	dnbinom	pnbinom	qnbinom	rnbinom
Normal	dnorm	pnorm	qnorm	rnorm
Poisson	dpois	ppois	qpois	rpois
Student's t	dt	pt	qt	rt
Uniform	dunif	punif	qunif	runif
Weibull	dweibull	pweibull	qweibull	rweibull

1.1.7 Probability Distributions

R includes the probability density function, the distribution function, the quantile function, and the random number generation for the main theoretical probability distributions which are shown in Table 1.4:

In the next chapters the beta, chi-squared, F, gamma, log-normal, and normal distribution mainly will be used. Let us analyze some basic examples.

The area under the normal distribution between −3 and 3 standard deviations is computed as:

```
> pnorm(3)- pnorm(−3)
[1] 0.9973
```

To generate a sample of size n = 15 from a gamma distribution with shape and scale parameter 1:

```
> set.seed(1234) # fixing the seed
> x <− rgamma(15,1,1); print(x)
[1] 0.011 0.747 0.786 0.117 0.922 0.176 1.437 0.157 0.220 3.528
[11] 0.063 0.147 0.599 0.213 0.504
```

1.1.8 Descriptive Statistics

The aim of the descriptive statistics is to summarize quantitative information about a dataset and usually is divided in:

- measures of central tendency
- measures of dispersion
- measures of shape

The measures of central tendency provide information about the central position of the data.

The most used of these measures is the arithmetic mean.

The arithmetic mean is the average of a group of observation and it is the preferred measure

$$\overline{x} = \sum_{i=1}^{n} x_i/n \tag{1.1}$$

where x_1, x_2,. . .,x_n are the observations and n the samples size.

The median is the value that divides the ranked data into two equal parts. In odd samples the median is the middle value while in even samples it is computed as the average of the two central values.

Odd samples:

$$Me = x_{n/2} \tag{1.2}$$

Even samples:

$$Me = \left(x_{n/2} + x_{n/2+1}\right)/2 \tag{1.3}$$

The mode is the most frequent occuring value. A dataset could have one, many, or neither mode.

The geometric mean: is another type of mean calculated as:

$$g = \left(\prod_{i=1}^{n} x_i\right)^{1/n} \tag{1.4}$$

The harmonic mean is a mean computed as:

$$h = \frac{n}{\sum_{i=1}^{n} 1/x_i} \tag{1.5}$$

The computation of these measures of central tendency is extremely easy. For instance

```
> mean(x)
[1] 0.6417593
> median(x)
[1] 0.219502
```

When the sample is small the mode can be selected visually ranking the data.

```
> sort(x)
[1] 0.011 0.063 0.117 0.147 0.157 0.176 0.213 0.220 0.504 0.599
[11] 0.747 0.786 0.922 1.437 3.528
```

As x was obtained via random number generation with eight decimal places and the sample size is only n = 15, it is practically impossible to get equal values. Therefore x does not have mode.

On the other hand the geometric and the harmonic mean respectively:

```
> prod(x) ^ (1 / length(x))
[1] 0.30
> 1 / mean(1 / x)
[1] 0.10
```

The measures of dispersion determine the deviation respect to the mean. The most commonly used are:

The variance that is the second central moment and is given by:

$$s^2 = \sum_{i=1}^{n} (x_i - \overline{x})/n - 1 \tag{1.6}$$

where $\overline{X}$ is the arithmetic mean.

The standard deviation is the most common measure and results in the square root of the variance.

$$s = \sqrt{\sum_{i=1}^{n} (x_i - \overline{x})/n - 1} \tag{1.7}$$

The range is the simplest measure.

$$R = x_{\max} - x_{\min} \tag{1.8}$$

The computation in R is as follows:

```
> sd(x)
[1] 0.89
> var(x)
[1] 0.80
```

The function range returns a vector with the minimum and maximum values. So, the range is the difference of these values.

```
> diff(range(x))
[1] 3.52
```

The measures of shape provide information about the shape and distribution of the data.

The skewness is an index that measures the asymmetry of the data. Negative values indicate the presence of tail on to the left and positive values to the opposite direction. It is given by:

$$g_1 = \frac{1}{n}\sum_{i=1}^{n}(x_i - \bar{x})^3 \bigg/ \left[\frac{1}{n}\sum_{i=1}^{n}(x_i - \bar{x})^2\right]^{3/2} \tag{1.9}$$

The kurtosis measures the peakedness of the distribution.

$$g_2 = \sum_{i=1}^{n}(x_i - \bar{x})^4/(n-1)S^2 \tag{1.10}$$

where S is the standard deviation.

Often the kurtosis "excess" is used due to the fact that in a normal distribution the kurtosis is equal to three. When this index takes negative values it is said that the distribution is platykurtic while positive values indicate leptokurtic distribution.

Histogram is a useful technique for assessing graphically the skewness and kurtosis.

R does not bring internal function to determine both skewness and kurtosis. However, they can be computed as follows:

```
> moments <- function(x,r){
> sum((x - mean(x)) ^ r) / length(x)}

> skew<-function(x){
> moments(x,3) / (moments(x,2) * sqrt(moments(x,2)))}

> kurtosis <- function(x) {
> moments(x,4) / (moments(x,2) * moments(x,2))}
```

Then:

```
> skew(x)
[1] 2.44
```

and

```
> kurtosis(x)
[1] 8.47
```

1.1.9 Statistical Inference (Hypothesis Testing)

Hypothesis testing is normally integrated by three parts: establishing of the hypotheses, calculation of the statistics, and computation of the *p*-value.

The simplest used ones are associated to the mean and variance comparison. For instance, the *t* test is employed to check if the mean is significant close to a target when the sample size n < 30.

Suppose we need to prove that the random number generated from a uniform distribution

$$H_o : \mu = 0.5$$
$$H_1 : \mu \neq 0.5$$

```
> set.seed(1234)
> x <- runif(20)
> t.test(x, mu = 0.5)
One Sample t-test
data: x
t = −0.47, df = 19, p-value = 0.64
alternative hypothesis: true mean is not equal to 0.5
95% confidence interval:
0.35 0.60
sample estimates:
mean of x
0.47
```

Being the *p*-value greater than the significance value α=0.05, the probability of Type I error is large. Therefore, there is no evidence to reject the null hypothesis (Ho). Besides, the test provides a 95% confidence interval to the mean.

Another hypothesis testing can be found by using apropos(".test")

1.1.10 A Short Introduction to Statistical Process Control (SPC). Univariate Control Charts

The introduction of the control chart dates back to the pioneer work of Walter A. Shewhart in 1920. It is based on the principle that in the normal distribution, 99.73% of the observations are between $\pm 3\sigma$.

A control chart is a graphical tool that allows to monitoring a quality characteristic through the time respect to a central line and an upper and lower control limit.

When one or more samples fall outside the control limits indicates the presence of a special cause; that is, a nonrandom shift has occurred. Consequently this assignable cause must be detected and eliminated.

When the process works without special causes, it is said that the process is in-control.

The $\overline{X}$ Chart which is the most studied and employed chart is based on the confidence interval for the mean

$$\overline{X} - Z_{\alpha/2}\sigma/\sqrt{n} \leq \mu \leq \overline{X} + Z_{\alpha/2}\sigma/\sqrt{n} \qquad (1.11)$$

With a probability of 1-α the mean will be in this interval. Za/2 it is usually substituted by 3 resulting

$$\overline{X} - 3\sigma/\sqrt{n} \leq \mu \leq \overline{X} + 3\sigma/\sqrt{n} \qquad (1.12)$$

Often in practice, the parameters μ and σ are unknown and must be estimated. Finally the chart results in

$$CL = \overline{\overline{X}} \quad UCL = \overline{\overline{X}} + A_2\overline{R} \quad LCL = \overline{\overline{X}} - A_2\overline{R} \qquad (1.13)$$

where

$$\overline{\overline{X}} = \sum_{k=1}^{m} \overline{X}_k/m \qquad (1.14)$$

$$\overline{X}_k = \sum_{i=1}^{n} X_i/n \qquad (1.15)$$

And

$$\overline{R} = \sum_{k=1}^{m} R_k/m \qquad (1.16)$$

being R_k=max(X_k)-min(X_k) (1.16) and A2 a constant selected according to the sample size.

The $\overline{X}$ Chart can also be computed using the standard deviation.

$$CL = \overline{\overline{X}} \quad UCL = \overline{\overline{X}} + A_3\overline{S} \quad LCL = \overline{\overline{X}} - A_3\overline{S} \qquad (1.17)$$

Normally the $\overline{X}$ chart is used jointly with a chart such as R and S chart to monitoring the process dispersion.

The R chart is as follows

$$CL = \overline{R} \quad UCL = D_4\overline{R} \quad LCL = D_3\overline{R} \qquad (1.18)$$

While the S chart

$$CL = \overline{S} \quad UCL = B_4\overline{S} \quad LCL = B_3\overline{S} \tag{1.19}$$

D_3, D_4, B_3, and B_4 are constants tabulated for the sample size.

In R the computation could be performed using the function qcc from the package named in the same way.

The construction of the chart is illustrated in the following example.

```
> library("qcc")
> set.seed(20)
fixing the seed of the generator.
> x <- round(rnorm(120,20,2),2)
> length <- matrix(x, ncol = 4, byrow = TRUE)
> par(mfrow = c(1,2))
> qcc(length, type = "xbar", std.dev = "RMSDF"); qcc(length, type = "R")
  (Fig. 1.2)
> qcc(length, type = "R")
```

1.1.11 Univariate Process Capability Indices (Cp, Cpk and Cpm)

Process capability can be conceived as the field in quality control focused on the determination of the feasibility by the process to fulfill with specifications.

Normally, the process capability is expressed in ratios or indices between tolerances and process performance. It is said that a process is capable when almost all of the samples are between the specifications limits.

Most capability studies consider normality, so the natural tolerance limits are placed 3σ above and below of the mean.

In literature many indices have been proposed to measure the capability, being the most recognized the following:

$$Cp = \frac{USL - LSL}{6\sigma} \tag{1.20}$$

$$Cpk = \min\left(\frac{USL - \mu}{3\sigma}, \frac{\mu - LSL}{3\sigma}\right) \tag{1.21}$$

$$Cpm = \frac{USL - LSL}{6\sqrt{\sigma^2 + (\mu - T)^2}} \tag{1.22}$$

$$T = \frac{USL - LSL}{2} \tag{1.23}$$

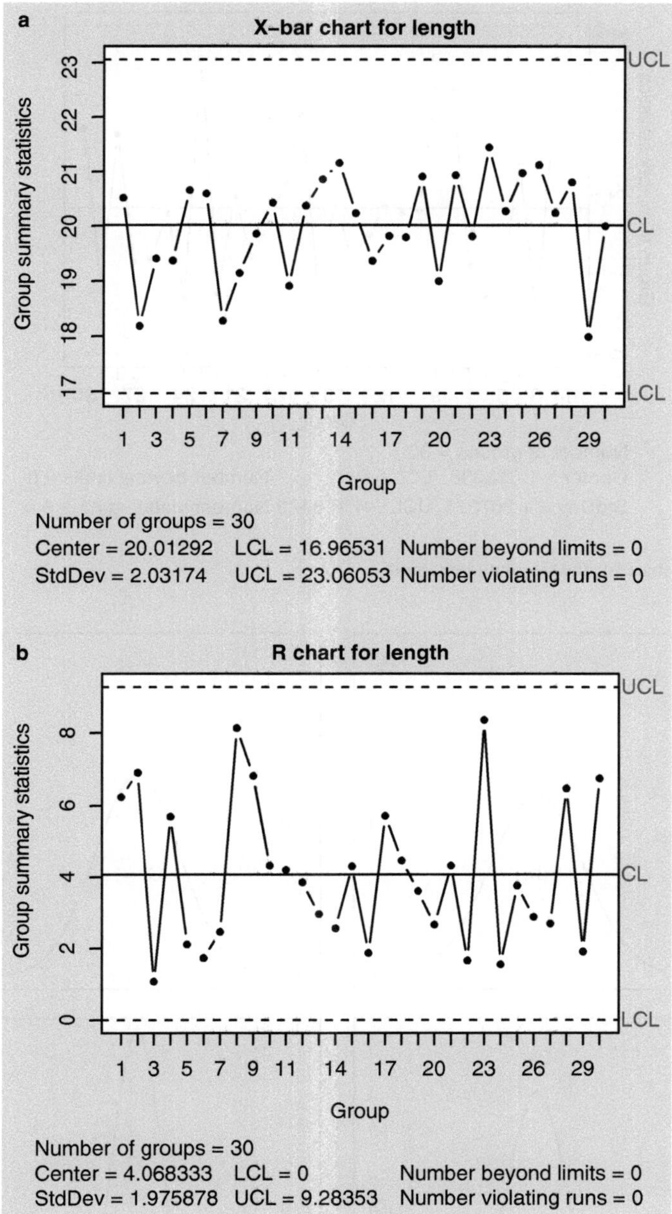

Fig. 1.2 (**a**) Xbar and (**b**) R chart for the simulated example

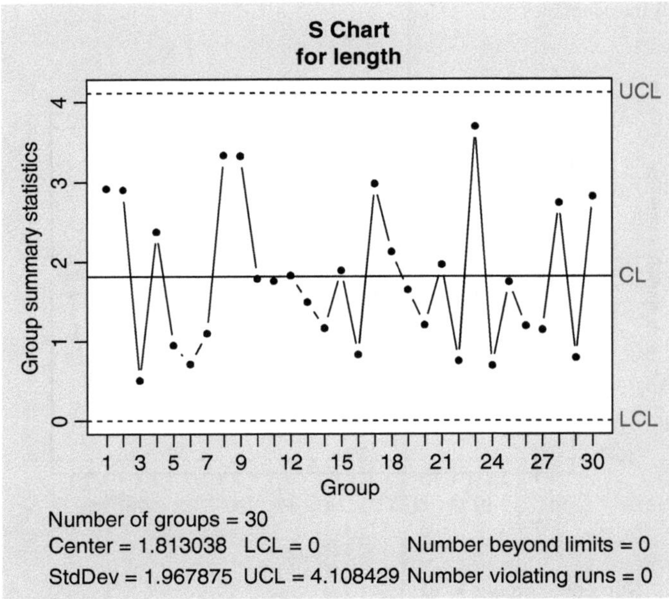

Fig. 1.3 S chart for the simulated example

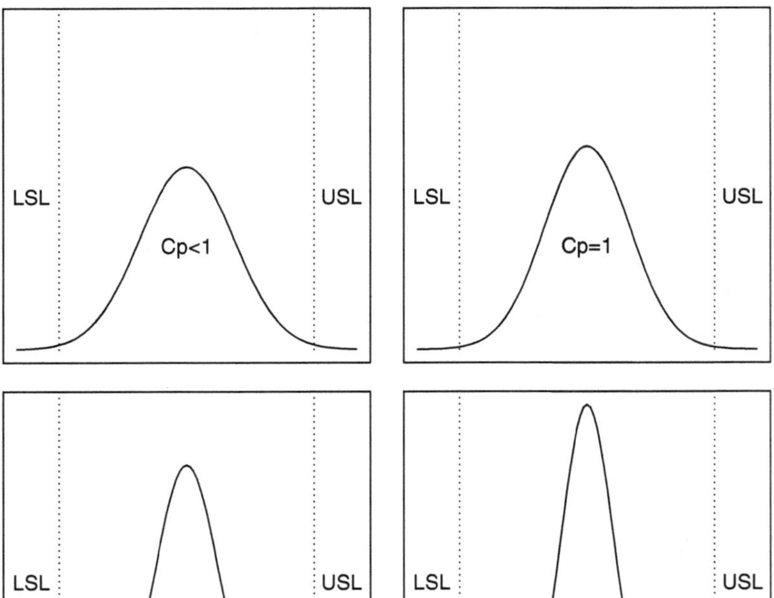

Fig. 1.4 Cp index for many process dispersions

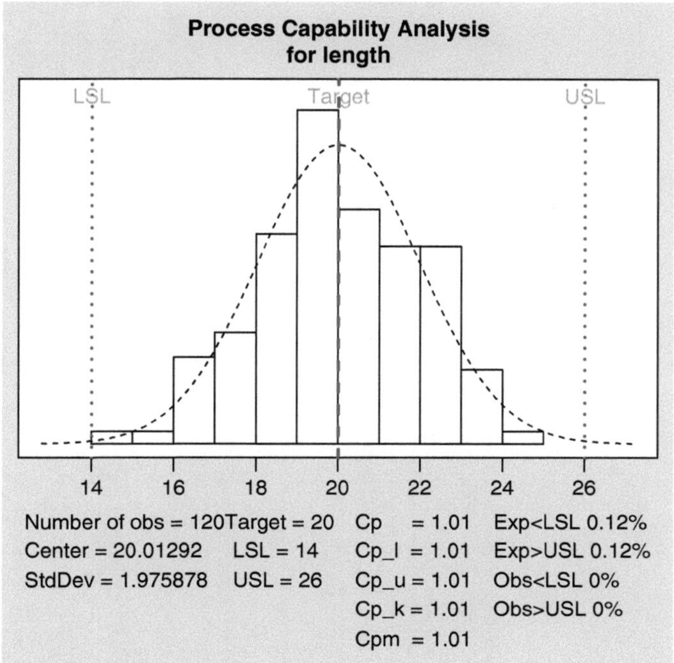

Fig. 1.5 Univariate capability indices for the simulated example

where USL and LSL are the upper and lower specification limits respectively and T the target. This last one is often fixed as the midpoint of specifications.

For more details see e.g.: (Kotz and Lovelace 1998) or (Montgomery 2004).

The parameters of the distribution are practically unknown and consequently σ must be replaced by S. In this case the term process performance is often used.

Figure 1.4 displays four possible scenarios for Cp. In all cases the process mean coincides with the target.

When $Cp = 1$ it is expected the 0.27% of nonconforming products. Whereas for values of 1.33 and 1.63, 64, and 1 ppm respectively.

Returning to the example about the length, the computation in R is as follows

```
> cap <- qcc(length, type = "xbar", nsigmas = 3, plot = FALSE)
> process.capability(cap, spec.limits = c(14,26)) (Fig. 1.5)
```

Since the indices Cp, Cpk, and Cpm are bigger than 1, consequently, the process is capable.

Chapter 2
Multivariate Control Charts

With the enhancements in data acquisition systems it is usual to deal with processes with more than one correlated quality characteristic to be monitored. A common practice is to control the stability of the process using univariate control charts. This practice increases the probability of false alarm of special cause of variation.

Therefore, the analysis should be performed through a multivariate approach; that is, the variables must be analyzed together, not independently.

In this chapter we present the multivariate normal distribution, the data structure of the multivariate problems dealt in this book, the mult.chart function that allows the computation in R, and the most used multivariate control charts:

- The control ellipsoid or χ^2 control chart
- The T^2 or Hotelling chart
- The Multivariate Exponentially Weighted Moving Average (MEWMA) chart
- The Multivariate Cumulative Sum (MCUSUM) chart
- The chart based on Principal Components Analysis (PCA)

2.1 The Multivariate Normal Distribution

The multivariate normal distribution (MVN) is the core of the multivariate statistical analysis. This is due to the fact that the sampling distributions of multivariate distributions exhibit approximately normality due to the central limit theorem.

In the univariate case if a random variable is normally distributed with mean μ and variance σ^2 it has a density function:

$$f(x) = \frac{1}{\sqrt{2\pi\sigma^2}} e^{-\frac{[(x-\mu)/\sigma]^2}{2}}, \tag{2.1}$$

where $-\infty < x < \infty$.

E. Santos-Fernández, *Multivariate Statistical Quality Control Using R*, SpringerBriefs in Statistics 14, DOI 10.1007/978-1-4614-5453-3_2,

The multivariate generalization is as follows. The upper part of exponent in the function can be written as

$$(x-\mu)^2/\sigma^2 = (x-\mu)(\sigma^2)^{-1}(x-\mu). \tag{2.2}$$

Since in multivariate normal distribution, the number of random variables is $(p) \geq 2$, then the generalization of (2.2) is

$$(x-\mu)'(\Sigma)^{-1}(x-\mu) \tag{2.3}$$

known as the Mahalanobis distance, where μ is the $p \times 1$ vector of expected values,

$$\mu' = \begin{bmatrix} \mu_1 & \mu_2 & \cdots & \mu_p \end{bmatrix} \tag{2.4}$$

and Σ the $p \times p$ variance–covariance matrix:

$$\Sigma = \begin{bmatrix} \sigma_{11} & \sigma_{12} & \cdots & \sigma_{1p} \\ \sigma_{21} & \sigma_{22} & \cdots & \sigma_{2p} \\ \vdots & \vdots & \ddots & \vdots \\ \sigma_{p1} & \sigma_{p2} & \cdots & \sigma_{pp} \end{bmatrix} \tag{2.5}$$

Finally, replacing in (2.1) the (2.2) by (2.3) and the constant $\frac{1}{\sqrt{2\pi\sigma^2}}$ by $\frac{1}{(2\pi)^{p/2}|\Sigma|^{1/2}}$ we have

$$f(x) = \frac{1}{(2\pi)^{p/2}|\Sigma|^{1/2}} e^{-\frac{(x-\mu)'(\Sigma)^{-1}(x-\mu)}{2}}, \tag{2.6}$$

where $-\infty < x_i < \infty$.

The notation used to denote a p-variate dataset with MVN is $N_p(\mu, \Sigma)$.

The bivariate case (p = 2 variables) is the most studied and applied in the practice. In this case the parameters of the distribution are given by the mean vector $\mu = \begin{bmatrix} \mu_1 \\ \mu_2 \end{bmatrix}$, and the covariance matrix $\Sigma = \begin{bmatrix} \sigma_{11} & \sigma_{12} \\ \sigma_{21} & \sigma_{22} \end{bmatrix}$.

The computation of the inverse of Σ results as follows:

$$\Sigma^{-1} = \frac{1}{\sigma_{11}\sigma_{22} - \sigma_{12}{}^2} \begin{bmatrix} \sigma_{22} & -\sigma_{12} \\ \sigma_{21} & \sigma_{11} \end{bmatrix}. \tag{2.7}$$

Replacing and standardizing into (2.6) it is relatively easy to achieve the density function:

$$f(x_1,x_2) = \frac{1}{2\pi\sqrt{\sigma_{11}\sigma_{22}(1-\rho_{12}^2)}} \times e^{\left\{-\frac{1}{2(1-\rho_{12}^2)}\left[\left(\frac{x_1-\mu_1}{\sqrt{\sigma_{11}}}\right)^2 + \left(\frac{x_2-\mu_2}{\sqrt{\sigma_{22}}}\right)^2 - 2\rho_{12}^2\left(\frac{x_1-\mu_1}{\sqrt{\sigma_{11}}}\right)\left(\frac{x_2-\mu_2}{\sqrt{\sigma_{22}}}\right)\right]\right\}}. \tag{2.8}$$

Example 2.1
In order to perform in R a graphical representation of a bivariate normal distribution with mean vector $\mu = \begin{bmatrix} 0 \\ 0 \end{bmatrix}$ and covariance matrix $\Sigma = \begin{bmatrix} 10 & 3 \\ 3 & 6 \end{bmatrix}$ we have

```
> mu <- c(0,0)
> sigma <- matrix(c(10,3,3,6),2,2)
> rho <- sigma[1,2] / (sqrt(sigma[1,1] * sigma[2,2]))
```

Defining the mean vector, the covariance matrix, and the correlation coefficient:

```
> var1 <- seq(-12,12,.7)
> var2 <- var1
> f <- matrix(0, length(var1), length(var1))
> for( i in 1:length(var1)){
>    for(j in 1:length(var1)){
>        f[i,j] <- 1/(2 * pi * sqrt(sigma[1,1] * sigma[2,2] * (1-rho ^ 2)))*exp(-1 /
  (2 * (1-rho ^ 2)) * ((var1[i] - mu[1]) ^ 2 / sigma[1,1] + (var2[j] - mu[2]) ^ 2 /
  sigma[2,2]-2 * rho * ((var1[i] - mu[1]) * (var2[j] - mu[2])) / (sqrt(sigma[1,1]) *
  sqrt(sigma[2,2]))))}}
> persp(var1, var2, f, xlab = "Variable 1", ylab = "Variable 2", zlab = "f(var1,
  var2)", theta = 30, phi = 30, r = 50, d = 0.2, expand = 0.6, ltheta = 90, lphi =
  180, nticks = 4)
```

Then R shows the bivariate density function (Fig. 2.1a).

Moreover it is possible to represent in a two-dimensional form using a contour plot (Fig. 2.1b):

```
> contour(var1, var2, f, xlab = "Variable 1", ylab = "Variable 2", nlevels = 8,
  drawlabels = F, xlim = c(-8,8), ylim = c(-8,8))
```

2.2 Data Structure

In order to provide a better comprehension in this section we offer a summary of the data structure and notation used for all methods.

As it is shown in Fig. 2.2, almost all the problems studied in this book deal with k samples of size n, taken from p quality characteristics or variables.

Where x_{ijk} is the ith observation of the jth quality characteristics on the kth sample.

Often the parameters of the distribution (μ and σ) are unknown and must be estimated through $\bar{\bar{x}}$ and S, respectively, which are computed as follows:

$$\bar{\bar{x}}_j = \frac{\sum_{k=1}^{m} \bar{x}_{jk}}{m}, \tag{2.9}$$

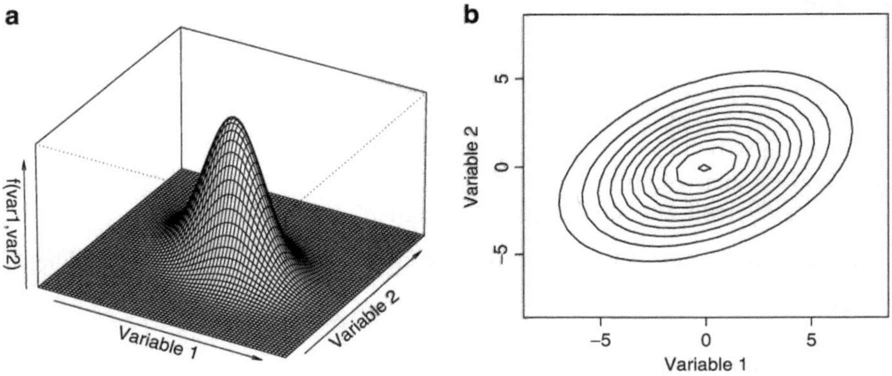

Fig. 2.1 (**a**) Bivariate density function. (**b**) Contour plot of a bivariate normal distribution

		Characteristic (j)			
		1	2	$\cdots$	p
	sample size (n)				
Sample (k)	1	$x_{111}\ x_{211}\ \cdots x_{n11}$	$x_{121}\ x_{221}\ \cdots x_{n21}$	$\cdots$	$x_{1p1}\ x_{2p1}\ \cdots x_{np1}$
	2	$x_{112}\ x_{212}\ \cdots x_{n12}$	$x_{122}\ x_{222}\ \cdots x_{n22}$	$\cdots$	$x_{1p2}\ x_{2p2}\ \cdots x_{np2}$
	$\vdots$	$\vdots$	$\vdots$	$\ddots$	$\vdots$
	m	$x_{11m}\ x_{21m}\ \cdots x_{n1m}$	$x_{12m}\ x_{22m}\ \cdots x_{n2m}$	$\cdots$	$x_{1pm}\ x_{2pm}\ \cdots x_{npm}$

Fig. 2.2 Graphical representation of the data structure

where

$$\bar{x}_{jk} = \frac{\sum_{i=1}^{n} x_{ijk}}{n}. \tag{2.10}$$

The case when the samples are composed by only one observation is called individual observations and will be studied in next sections.

On the other hand, S is estimated as

$$S = \begin{matrix} \bar{S}_1^2 & \bar{S}_{12} & \cdots & \bar{S}_{1p} \\ \bar{S}_{12} & \bar{S}_2^2 & \cdots & \bar{S}_{2p} \\ \vdots & \vdots & \ddots & \vdots \\ \bar{S}_{1p} & \bar{S}_{2p} & \cdots & \bar{S}_2^2 \end{matrix}, \tag{2.11}$$

where the diagonal elements are variances associated to the characteristics p and the non-diagonal are the covariances. Being

$$\bar{S}_j^2 = \frac{\sum_{k=1}^{m} S_{jk}^2}{m} \tag{2.12}$$

with

$$S_{jk}^2 = \frac{\sum_{i=1}^{n} \left(x_{ijk} - \bar{x}_{jk}\right)^2}{n-1} \tag{2.13}$$

and

$$\bar{S}_{jl} = \frac{\sum_{k=1}^{m} S_{jlk}}{m} \tag{2.14}$$

with $j \neq l$ being

$$S_{jlk} = \frac{\sum_{i=1}^{n} \left(x_{ijk} - \bar{x}_{jk}\right)\left(x_{ijk} - \bar{x}_{lk}\right)}{n-1}. \tag{2.15}$$

The mean vector (Xmv) is obtained in R as x.jk <− apply(3D.array, 1:2, mean).

First calculating the mean of each sample (see (2.10)), and then using the colMeans function (see (2.9)):

Xmv <− colMeans(x.jk)

With respect to the sample covariance matrix, it can be achieved directly using the function covariance included in MSQC package:

S <− covariance(x)

2.3 The mult.chart Function

The performing of the multivariate control chart in R can be carried out with the function mult.chart which is a general function that allows to compute the most accepted and diversified continuous multivariate chart such as

- χ^2
- Hotelling T^2
- MEWMA
- MCUSUM according to Crosier (1988)
- MCUSUM by Pignatiello and Runger (1990)

The selection of the chart to use is done by specifying the argument type = "t2," "mewma," "mcusum," or "mcusum2" in the same order previously introduced.

For more details about the function see the package manual:

```
> help(package = "MSQC")
```

In the function x must be a matrix or an array and jointly with type are the only compulsory arguments.

Other important functionalities are the Phase that can be I or II (being I for default) and the significance level (alpha) fixed in 0.01.

As it is shown in the next section, the covariance matrix (S) and mean vector (Xmv) can be entered to be used in Phase II.

Finally the function mult.chart returns:

- The T^2 statistics
- The Upper Control Limit (UCL)
- The sample covariance matrix (S)
- The mean vector (Xmv)
- And if any point falls outside of the UCL and its decomposition

The execution of the function takes few hundredth of a second as can be tested by

```
> system.time(mult.chart(dowel1, type = "chi", alpha = 0.05))
```

2.4 Contour Plot and χ^2 Control Chart

In multivariate normal distribution the density is described by an ellipsoid centered at mean vector with axes in direction to the eigenvectors (e) of the covariance matrix, setting μ as the origin and with length

$$\pm c\sqrt{\lambda_j}e_j \tag{2.16}$$

being

$$(x-\mu)'\Sigma^{-1}(x-\mu) = c^2. \tag{2.17}$$

If x follows $N_p(\mu, \Sigma)$ then $(x-\mu)'(\Sigma)^{-1}(x-\mu)$ is $\chi^2_{\alpha,p}$. Therefore,

$$(x-\mu)'\Sigma^{-1}(x-\mu) \leq \chi^2_{\alpha,p}. \tag{2.18}$$

Example 2.2
To illustrate the construction of an ellipsoid contour consider the dataset called dowel that comprises 40 samples from two correlated quality characteristics (diameter and length) collected from a manufacturing process of a dowel pin.

To call the dataset, just use

```
> data("dowel1")
```

The construction of the control ellipse for dowel1 results as follows. Setting the significance level:

```
> alpha <- 0.05 and
> p <- ncol(dowel1)
```

Then the mean vector and the covariance matrix are estimated:

```
> Xmv <- colMeans(dowel1)
```

The function colMeans was used directly due to the fact that this is a problem of individual observations:

```
> S <- covariance(dowel1)
```

So we have

$\mu' = [0.50 \quad 1.00]$ and $\Sigma = \begin{bmatrix} 4.91e-05 & 8.58e-05 \\ 8.58e-05 & 4.20e-04 \end{bmatrix}$.

The computation of the eigenvalues and eigenvectors is based on the R function eigen:

```
> DDe <- eigen(S)$values
> Ue <- eigen(S)$vectors
```

For more details see help function.

Then we have

$\lambda' = [4.39e-04 \quad 3.02e-05]$, $e_1' = [0.22 \quad -0.98]$, and $e_2' = [-0.98 \quad 0.22]$.

Plotting the ellipsoid origin given by Xmv. (at 0.50, 1.00) with the respective axes labels and ranges:

```
> plot(Xmv[1], Xmv[2], xlim = c(0.46,0.54), ylim = c(0.95,1.06), xlab = "diameter", ylab = "length",pch = 3)
```

The direction of the ellipsoid axes is given by the eigenvectors:

```
> inc <- atan ((Xmv[2] + Ue[2,1] - Xmv[2]) / (Xmv[1] + Ue[1,1] - Xmv[1]))
```

Then we must compute the lengths regarding the x- and y-axes as follows:

```
> b <- (sqrt(DDe[1]) * sqrt(qchisq(1 - alpha,p))) * sin(inc)
> a <- (sqrt(DDe[1]) * sqrt(qchisq(1 - alpha,p))) * cos(inc)
> d <- (sqrt(DDe[2]) * sqrt(qchisq(1 - alpha,p))) * sin(inc)
> c <- (sqrt(DDe[2]) * sqrt(qchisq(1 - alpha,p))) * cos(inc)
```

Finally, we trace the axes using

```
> arrows(Xmv[1], Xmv[2], Xmv[1] + a, Xmv[2] + b)
> arrows(Xmv[1], Xmv[2], Xmv[1] - a, Xmv[2] - b)
> arrows(Xmv[1], Xmv[2], Xmv[1] - d, Xmv[2] + c)
> arrows(Xmv[1], Xmv[2], Xmv[1] + d, Xmv[2] - c)
```

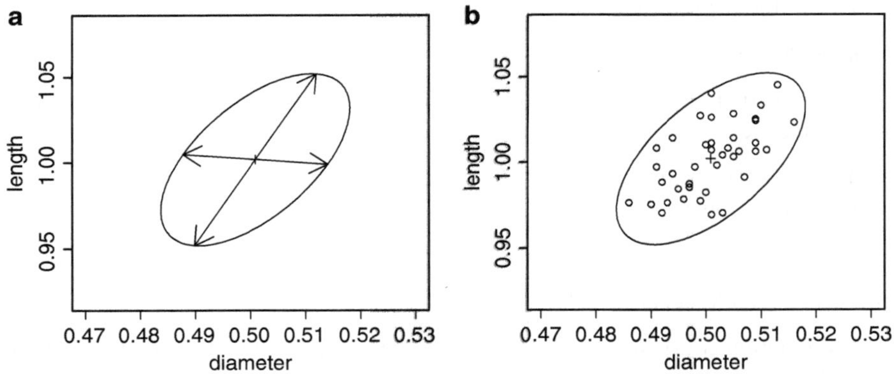

Fig. 2.3 (**a**) Confidence ellipse with the axes for the dowel1 dataset. (**b**) Scatterplot for the dowel1 dataset with the confidence ellipse

The ellipse results by connecting the axes extremes.

Fortunately it is relatively easy to draw an ellipse in R, making use of this algorithm:

```
> angle <- seq(0, 2 * pi, length.out = 200)
> ch <- cbind(sqrt(qchisq(1 - alpha,2)) * cos(angle), sqrt(qchisq(1 - alpha,2)) *
  sin(angle))
> lines(t(Xmv - ((Ue %*% diag(sqrt(DDe))) %*% t(ch))),type = "l")
```

Figure 2.3a shows the result.

This procedure is known as confidence ellipsoid. Figure 2.3b shows the addition of the points of the dowel1 array:

```
> points(dowel1)
```

Obtaining no points outside the ellipse, there is no evidence of special causes; therefore the process is in-control. Notice that if the limits from the univariate individual control chart are plotted, how much this area differs to the confidence ellipse. In fact, four points fall outside to this area (Fig. 2.4).

The difficulty to identify the points beyond the confidence ellipsoid is one of the main drawbacks of the tool, although it can be solved by inserting the sample number in plot when the amount of points is not large.

Another disadvantage is the complexity to construct the ellipsoid when $p > 2$ which can be solved using the χ^2 control chart that results by plotting the test statistics:

$$n(x - \mu)'(\Sigma)^{-1}(x - \mu) = \chi^2_{\alpha,p}, \tag{2.19}$$

where n is the sample size and the upper control limit:

$$UCL = \chi^2_{\alpha,p}. \tag{2.20}$$

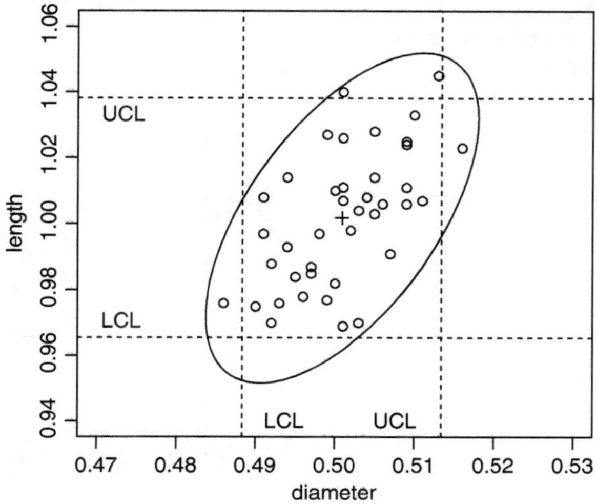

Fig. 2.4 Scatterplot for the dowel1 dataset with the confidence ellipse and the Shewhart control limits

```
[1] "Chi-squared Control Chart"
$ucl
[1] 6
$t2
[1,] 1.61
 [2,] 0.30
...
 [39,] 1.58
[40,] 1.64
$Xmv
[1] 0.50 1.00
$covariance
          [,1]        [,2]
[1,] 4.91e -05 8.59e-05
[2,] 8.59e -05 4.20e-04
```

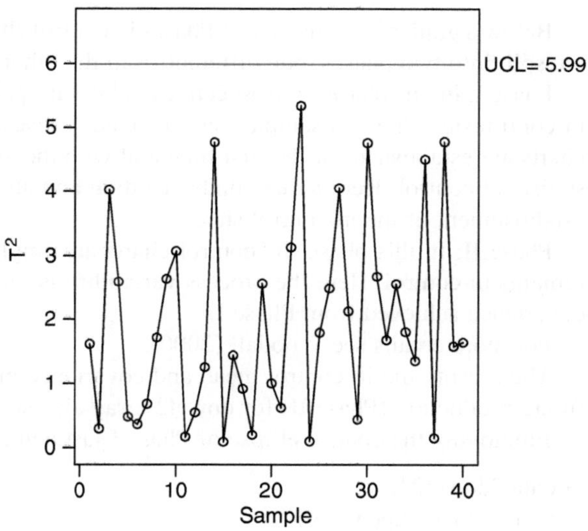

Fig. 2.5 χ^2 control chart for the dowel1 dataset

When μ and Σ are estimated through a sufficiently large sample then the χ^2 chart can be used although the parameters are unknown.

Through the function mult.chart

```
> mult.chart(dowel1, type = "chi", alpha = 0.05)
```

The function returns (Fig. 2.5):

Showing results alike to the control ellipsoid. An advantage of this chart is that it allows the evolution of the samples along time.

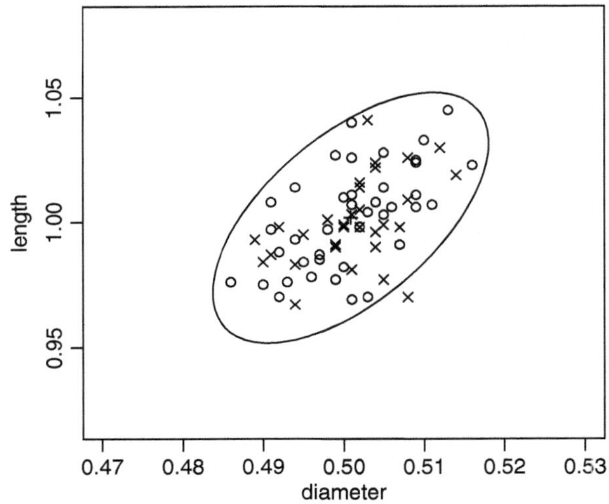

Fig. 2.6 Phase II confidence ellipsoid for the dowel2 dataset

Below a guidance on the use of Phases in control charts is given. Usually, studies are split into two phases, one different from the other.

Phase I: In this phase a retrospective analysis is applied to assess if the process is in-control since the first sample was collected. These studies are used when control charts are established for the first time and with the aim of bringing the process to statistical control. Here a deep understanding and analysis are required before the establishment of the in-control state.

Phase II: In this phase the control charts are employed to verify if the process remains in-control. Here the process variability is monitored using the mean and covariance achieved from Phase I.

For more details see Woodall (2000).

Then, using the in-control mean and covariance matrix it is possible to control future production (Phase II) for dowel2 array also stored in the MSQC package.

Employing the control ellipse of Phase I just add the Phase II points as

```
> data("dowel2")
> points(dowel2,pch = 4)
```

The argument pch = 4 allows to differentiate the points. One point falls outside the 95th confidence ellipsoid, indicating the presence of special cause in the process (Fig. 2.6).

Conversely the χ^2 control chart can be used.

The mean vector and covariance matrix of the in-control Phase I process are used as the parameters of the distribution:

```
> vec <– (mult.chart(dowel1, type = "chi", alpha = 0.05)$Xmv)
> mat <– (mult.chart(dowel1, type = "chi", alpha = 0.05)$covariance)
```

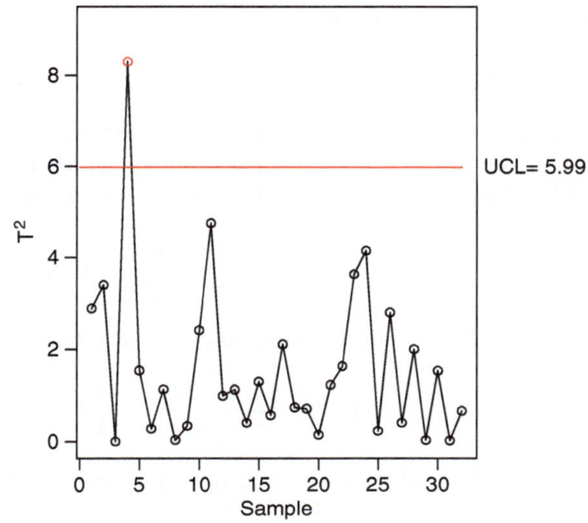

Fig. 2.7 χ^2 control chart in Phase II for the dowel2 dataset

Finally they are passed in the function mult.chart:

> mult.chart(dowel2, type = "chi", Xmv = vec, S = mat, alpha = 0.05)

The fourth sample falls beyond the UCL; as a consequence, there is evidence of special causes, and then the process is out-of-control (Fig. 2.7).

2.5 Hotelling T² Control Chart (Phase I)

The origin of the T^2 control chart dates back to the pioneer works of Harold Hotelling who applied this method to the bombsight problem in Second World War. The Hotelling (1947) procedure has become without doubt the most applied in multivariate process control and it is the multivariate analogous of the Shewhart control chart. For that reason, it is also known as multivariate Shewhart control chart.

Often in practice the parameters μ and Σ are unknown and consequently must be estimated across the unbiased estimators $\bar{x}$ and S. Based on the multivariate generalization of the t statistic from univariate normal theory:

$$t = \frac{\bar{x} - \mu}{S/\sqrt{n}} \tag{2.21}$$

making

$$t^2 = \frac{(\bar{x} - \mu)^2}{S^2/n} = n(\bar{x} - \mu)\left(S^2\right)^{-1}(\bar{x} - \mu), \tag{2.22}$$

so the generalization results in

$$T^2 = n(\overline{X} - \overline{\overline{X}})'(S)^{-1}(\overline{X} - \overline{\overline{X}}) \tag{2.23}$$

with $\bar{X}$ and S being the vector of means and the covariance matrix, respectively.

The statistics T^2 follows an F distribution with p and (mn − m − p + 1) degrees of freedom. Therefore for establishing the control in Phase I the UCL results in

$$UCL = \frac{p(m-1)(n-1)}{mn - m - p + 1} F_{\alpha, p, mn-m-p+1}. \tag{2.24}$$

While for monitoring future observations (Phase II) the limit is given by

$$UCL = \frac{p(m+1)(n-1)}{mn - m - p + 1} F_{\alpha, p, mn-m-p+1}. \tag{2.25}$$

Here, (2.25), the number of samples (m) refers to the preliminary samples taken to establish the in-control state (Phase I). Notice that this chart lacks lower control limits (LCL) analogously to the χ^2 chart.

This chart is employed in introductory multivariate studies and has a good performance in detection of large shifts in the mean.

According to Lowry and Montgomery (1995) the application of this chart requires a number of quality characteristics between 2 and 10, taking more than 20 samples (often more than 50) of size 2, 3, or 10. These values are sometimes limited by the very nature of the problem, though.

The following example explains the construction of this chart.

Example 2.3

In the manufacturing process of a specific carbon fiber tubing three correlated quality characteristics are measured: inner diameter, thickness, and length of the tubes in inches. The dataset named carbon1 contains the information of 30 samples of size 8 taken and summarized in Table 2.1.

The sample mean vector, sample covariance, and correlation matrix result as follows:

$$\bar{x} = \begin{bmatrix} 0.99 \\ 1.04 \\ 49.98 \end{bmatrix}; \quad S \times 100 = \begin{bmatrix} 0.25 & 0.36 & 0.67 \\ 0.36 & 1.45 & 1.02 \\ 0.67 & 1.02 & 5.92 \end{bmatrix}; \quad r = \begin{bmatrix} 1 & 0.63 & 0.57 \\ 0.63 & 1 & 0.38 \\ 0.57 & 0.38 & 1 \end{bmatrix}.$$

It can be easily appreciated the direct correlation among the variables; being significant between the inner diameter with the others.

As we are in the presence of a trivariate process, it is possible a spatial representation. Figure 2.8 shows the three-dimensional scatterplot with the 99% ellipsoid. All the points of the swarm are inside the ellipsoid.

A scatterplot matrix is presented below and corroborates the information offered by the correlation matrix about the direct correlation between variables (Fig. 2.9)):

```
pairs(carbon1,labels=c("inner diameter", "thickness", "length"))
```

Table 2.1 Carbon fiber data

	Subgroup mean			Variance (×100)			Covariance (×100)			
Sample	Inner (X_1)	Thickness (X_2)	Length (X_3)	S_{21}	S_{22}	S_{23}	S_{12}	S_{13}	S_{23}	T^2
1	1.03	1.08	50.16	0.15	1.19	2.98	−0.08	0.40	−0.42	4.99
2	0.97	0.95	49.92	0.40	1.01	6.38	0.16	1.03	0.25	4.66
3	1.01	1.05	50.14	0.17	0.99	1.96	0.31	0.38	0.81	3.28
4	1.00	1.05	49.91	0.26	1.14	5.73	0.36	−0.29	0.82	1.93
5	0.96	1.00	49.83	0.43	3.25	12.37	0.92	1.91	4.61	5.62
6	1.03	1.07	50.05	0.30	1.52	1.69	0.43	0.30	−0.48	4.64
7	0.96	1.02	49.95	0.17	0.58	4.34	0.03	0.34	0.74	5.50
8	1.00	1.02	50.02	0.17	0.79	6.16	0.11	−0.11	0.32	0.87
9	1.00	1.10	50.03	0.20	1.43	1.87	0.34	−0.07	−0.88	2.87
10	0.99	1.02	50.00	0.13	0.53	6.58	0.11	0.18	−0.13	0.49
11	1.01	1.10	50.01	0.18	1.31	3.41	0.11	0.19	0.36	2.40
12	1.02	1.07	49.99	0.24	0.81	3.41	0.05	0.70	0.67	1.98
13	0.97	1.00	49.96	0.48	2.36	17.72	0.98	2.44	5.17	2.36
14	1.01	1.05	50.04	0.13	1.08	7.20	0.18	0.16	1.98	0.96
15	1.00	1.06	50.02	0.24	1.14	7.80	0.26	1.01	0.43	0.35
16	1.00	1.03	49.99	0.39	1.66	3.69	0.71	0.98	1.27	0.22
17	1.00	1.04	49.99	0.10	1.27	7.71	0.00	0.44	−2.09	0.05
18	0.98	1.00	49.94	0.18	1.56	5.40	0.26	0.85	2.22	0.86
19	0.98	0.96	49.93	0.24	1.61	5.68	0.55	0.18	0.55	3.43
20	1.01	1.07	50.02	0.37	2.55	4.91	0.64	1.16	3.33	1.08
21	0.98	1.03	49.96	0.28	0.39	7.21	0.15	1.39	0.64	0.45
22	0.99	1.04	50.07	0.23	2.46	8.24	0.60	0.70	1.74	2.74
23	0.95	0.92	49.86	0.41	1.82	2.69	0.73	0.40	0.32	9.43
24	1.00	1.09	50.05	0.15	0.75	9.27	0.12	0.69	−0.29	2.93
25	0.99	1.01	49.96	0.51	1.87	7.08	0.56	1.56	1.63	0.46
26	0.99	1.02	49.89	0.12	0.75	7.04	0.19	0.59	1.34	1.34
27	0.99	1.03	49.84	0.24	3.80	7.47	0.72	0.87	2.20	3.39
28	1.01	1.04	49.97	0.06	0.80	2.46	0.14	0.08	0.05	1.97
29	1.03	1.10	50.07	0.19	1.29	2.38	0.43	0.36	0.72	3.54
30	1.01	1.08	49.97	0.33	1.75	6.78	0.69	1.27	2.73	1.40

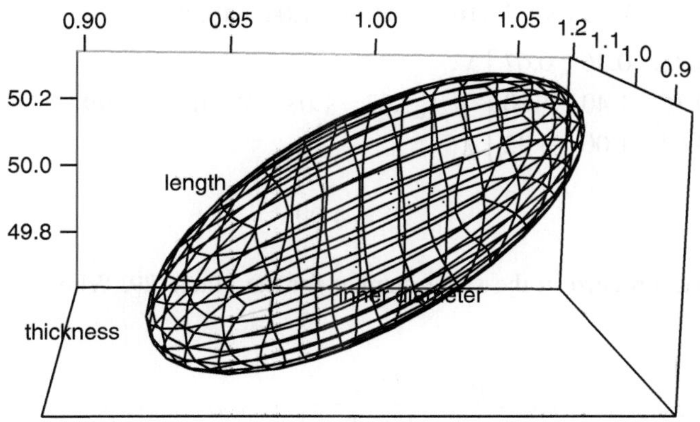

Fig. 2.8 3D scatterplot with the 99% confidence region

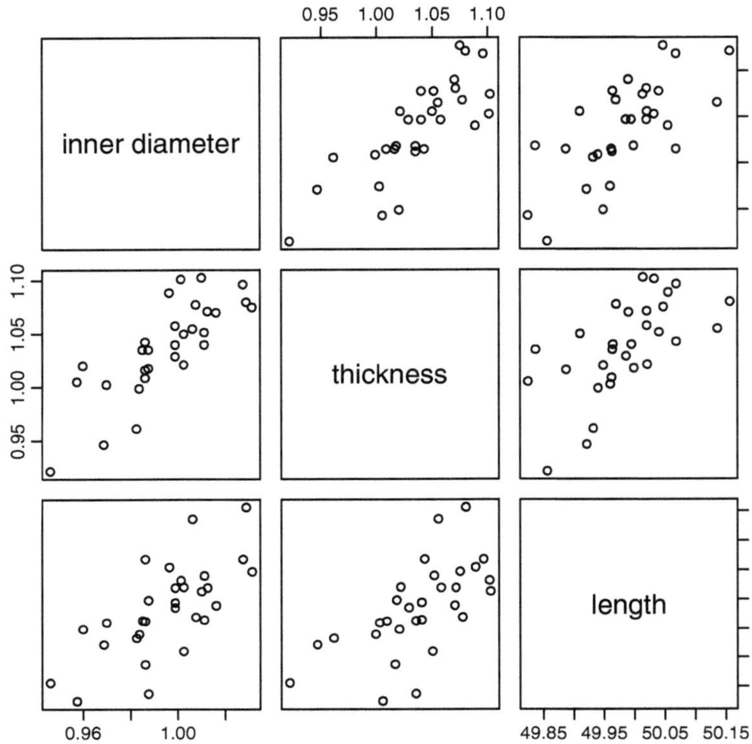

Fig. 2.9 Matrix of scatterplot

After this explanatory analysis let us compute the T^2 statistics:

$$T^2 = n(\bar{X} - \bar{\bar{X}})'(S)^{-1}(\bar{X} - \bar{\bar{X}})$$

$$T_1^2 = 8 \times \{[1.03 \quad 1.08 \quad 50.16] - [0.99 \quad 1.04 \quad 49.98]\}' \times \left(\begin{bmatrix} 0.25 & 0.36 & 0.67 \\ 0.36 & 1.40 & 1.00 \\ 0.67 & 1.00 & 5.92 \end{bmatrix}\right)^{-1} \{[1.03 \quad 1.08 \quad 50.16] - [0.99 \quad 1.04 \quad 49.98]\}$$

$$T_1^2 = 4.99.$$

After that, proceed in the same manner for the others 29. Whereas the limit is computed as

$$UCL = \frac{p(m-1)(n-1)}{mn - m - p + 1} F_{\alpha,\, p,\, mn-m-p+1}$$

```
[1] "Hotelling Control Chart"
$ucl
[1] 11.35
$t2
     [,1]
 [1,] 4.99
 [2,] 4.66
...
[29,] 3.54
[30,] 1.40
$Xmv
[1]  0.99  1.04 49.98
$covariance
       [,1]   [,2]   [,3]
[1,] 0.0025 0.0036 0.0067
[2,] 0.0036 0.0140 0.0100
[3,] 0.0067 0.0100 0.0590
```

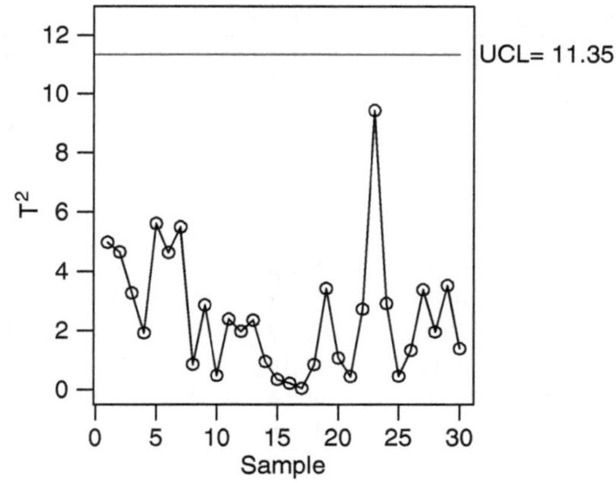

Fig. 2.10 Hotelling control chart of the carbon1 dataset

$$UCL = \frac{3(30-1)(8-1)}{30*8-30-3+1} F_{0.01,3,30*8-30-3+1} = \frac{609}{208} F_{0.01,3,30*8-30-3+1} = 11.35.$$

To perform this computation in R we will use the dataset called carbon1:

```
> data("carbon1")
> mult.chart(type = "t2", carbon1)
```

The output is shown in (Fig. 2.10).

Notice that no points fall beyond the UCL; therefore, the process is in statistical control. In order to work with any object of the function output just use the $ operator. For instance, to acquire only the T^2 statistics type

```
> mult.chart(type = "t2", carbon1)$t2
```

2.6 Interpretation, Decomposition, and Phase II

In control chart when one or more points fall outside of control limits then there is evidence that the process has suffered a nonrandom shift.

In univariate alternative the statistics proceeds from only one variable, but in multivariate problems the identification of the source that causes the out-of-control signal is more complex.

Table 2.2 Carbon fiber data of the Phase II

	Subgroup means					Subgroup means			
Sample	Inner (X_1)	Thickness (X_2)	Length (X_3)	T^2	Sample	Inner (X_1)	Thickness (X_2)	Length (X_3)	T^2
1	1.01	1.07	49.88	4.84	14	1.04	1.07	50.14	6.64
2	1.00	1.01	49.93	1.49	15	1.02	1.08	50.11	2.73
3	1.00	1.03	49.96	0.33	16	0.99	1.04	50.11	4.58
4	1.02	1.19	50.15	14.19	17	1.02	1.06	50.03	2.64
5	1.01	0.99	50.03	4.68	18	0.98	1.04	49.89	2.17
6	1.01	1.04	50.02	0.68	19	0.99	1.02	49.80	5.51
7	1.02	1.03	50.17	6.49	20	1.03	1.05	50.13	6.79
8	0.99	1.06	50.06	3.27	21	1.00	1.08	50.06	1.72
9	1.01	1.04	49.98	1.63	22	1.03	1.07	50.20	6.52
10	0.99	1.03	49.92	0.65	23	1.00	1.05	50.04	0.81
11	0.98	1.05	49.95	1.27	24	1.01	1.04	49.93	3.02

The issue usually named *decomposition* determines which variables are responsible for the variation when a nonrandom signal occurs. A frequent practice consists in performing a univariate chart although this analysis is often inefficient.

In the same way Alt (1985) proposed the use of Bonferroni control limits. After that, this field has been widely investigated. See for instance Murphy (1987), Doganaksoy et al. (1991), Wierda (1994), etc. The method suggested by Mason et al. (1995) is the most widely accepted to face the decomposition, though.

Example 2.4
Return to the carbon data to illustrate a decomposition technique.

Being the process in control in the previous Example 3.3 the mean vector and the covariance matrix from Phase I were used to monitor the process in future production (Phase II). 25 samples of size n = 8 were obtained, which are summarized in Table 2.2.

The computation of the statistics using the in-control sample mean and covariance matrix is the following:

$$T_1^2 = 8 * \{[1.01 \quad 1.07 \quad 49.88] - [0.99 \quad 1.04 \quad 49.98]\}' \times$$
$$\left(\begin{bmatrix} 0.25 & 0.36 & 0.67 \\ 0.36 & 1.45 & 1.02 \\ 0.67 & 1.02 & 5.92 \end{bmatrix} /100\right)^{-1} \times \{[1.01 \quad 1.07 \quad 49.88] - [0.99 \quad 1.04 \quad 49.98]\},$$

$T_1^2 = 4.84$, and so forth for the others.
The UCL in Phase II results in

$$UCL = \frac{p(m+1)(n-1)}{mn - m - p + 1} F_{\alpha,p,mn-m-p+1}$$

$$UCL = \frac{3(30+1)(8-1)}{30 \times 8 - 30 - 3 + 1} F_{0.01,3,30\times8-30-3+1} = \frac{651}{208} F_{0.01,3,30\times8-30-3+1} = 12.13.$$

Analyzing the T^2 values in Table 2.2 it is easy to determine that the fourth sample falls beyond UCL. This is evidence that a shift took place.

The method also called MYT decomposition (Mason et al. 1995) deals with the identification of the contribution of each individual variable and all the possible combinations increasing the group size. The scheme they proposed can be described as follows:

1. Compute the T^2 statistics (each variable independently)

$$T_j^2 = \frac{n\left(x_j - \bar{x}_j\right)^2}{S_j^2} \tag{2.26}$$

   where $\bar{x}_j$ and S_j^2 are the mean and the variance of the jth variable.
2. Compare with their corresponding threshold according to the phase and the sample size. For instance for individual observations in Phase II:

$$UCL = \frac{p(m+1)(m-1)}{m(m-p)} F_{\alpha,p,m-p}. \tag{2.27}$$

3. Then exclude the variables that satisfy:

$$T_j^2 > UCL. \tag{2.28}$$

4. Construct the T^2 statistics for the combinations of the remaining variables; e.g.: $T^2_{(x_1,x_2)}$.
5. Exclude variables whose T^2 exceed the limits. For p = 2 the limits are:

$$UCL = \frac{2(m+1)(m-1)}{m(m-2)} F_{\alpha,2,m-2}. \tag{2.29}$$

6. Carry out iteratively until the last combination that includes all the quality characteristics.

Returning to the example, the decomposition of the 4th sample using this methodology results in

$$\bar{X}_{4,j} = [\,1.02 \quad 1.19 \quad 50.15\,]; \quad T_1^2 = 2.44; \quad T_2^2 = 12.26; \quad T_3^2 = 3.82$$

$$UCL = \frac{1(30+1)(30-1)}{30(30-1)} F_{0.01,1,30-1} = 7.85.$$

Fig. 2.11 Hotelling control chart of the carbon2 dataset in Phase II

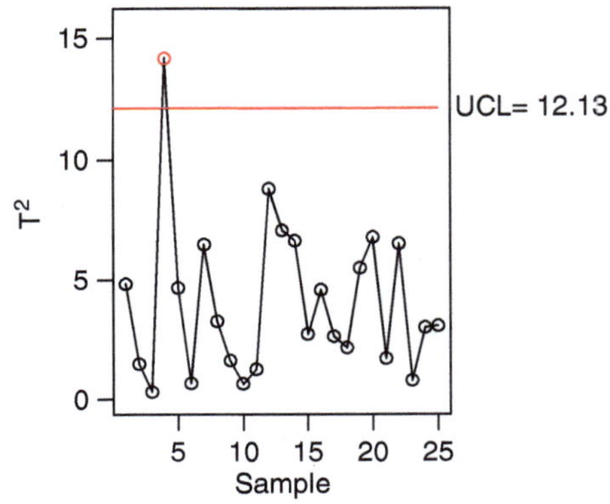

As $T_2^2 = 12.26 > 7.85$, x_2 contributes significantly to the out-of-control signal, and consequently it is excluded.

All combinations that include x_2, i.e.: $T^2_{(x_1,x_2)}$, $T_{(x_2,x_3)})^2$, and $T^2_{(x_1,x_2,x_3)}$ are eliminated. Only the pair $T^2_{(x_1,x_3)}$ remains to be checked:

$$T^2_{(x_1,x_3)} = 8 * \{[\,1.02 \quad 50.15\,] - [\,0.99 \quad 1.04\,]\}' * \begin{bmatrix} 0.0025 & 0.0067 \\ 0.0067 & 0.0592 \end{bmatrix}^{-1} * \{[\,1.02 \quad 50.15\,] - [\,0.99 \quad 1.04\,]\} = 4.15$$

The UCL associated is

$$UCL = \frac{2(30+1)(30-1)}{30(30-2)} F_{0.01,2,30-2} = 9.77.$$

As $T^2_{(x_1,x_3)}$ does not exceed the UCL the combination of x_1 and x_3 does not contribute to the signal.

In order to compute in R the T^2 chart with the new 25 samples in Phase II it is necessary to use the values of the mean vector, the covariance matrix, and their sample size stored in colm:

```
> Xmv <- mult.chart(carbon1, type = "t2") $Xmv
> S <- mult.chart(carbon1, type = "t2") $covariance
> colm<-nrow(carbon1)
then
> data("carbon2")
> mult.chart(carbon2, type = "t2", Xmv = Xmv, S = S, colm = colm)
```

The results are presented above (Fig. 2.11).

The process is out-of-control since the fourth sample falls outside the UCL. When this happens the mult.chart function returns a table with the T^2 value of the decomposition, the UCL, and the p-value for all possible combinations of variables.

The following point(s) falls outside the control limits[1] 4:
$'Decomposition of'
[1] 4

t^2	Decomp	UCL	p-Value	1	2	3
[1,]	3.3800	6.9823	0.0763	1	0	0
[2,]	12.2223	6.9823	0.0015	2	0	0
[3,]	4.0347	6.9823	0.0540	3	0	0
[4,]	12.3549	9.7767	0.0001	1	2	0
[5,]	4.8015	9.7767	0.0158	1	3	0
[6,]	12.9364	9.7767	0.0001	2	3	0
[7,]	13.6477	12.1347	0.0000	1	2	3

The first three rows present for each of the quality characteristic analyzed (decomposed individually). The x_2 represents the source of variability since p-value = 0.0015. Obviously all the combinations that include x_2 exceed their respective value of UCL. Finally, the same results are obtained.

2.6.1 T^2 for Individuals

In the previous section we have studied rational subgroup cases in which each sample is composed by more than one observation.

However, in many processes, due to its own nature, it can only measure one observation at each time interval. This case is frequently named for individuals.

It means that in data structure of the process only one observation per variable is recorded at the time m therefore, n = 1.

In this case T^2 bears only few modifications:

$$T^2 = (X - \overline{X})'(S)^{-1}(X - \overline{X}) \tag{2.30}$$

and evidently the control limits must be modified due to the absence of n. In this case, Tracy et al. (1992) propose for Phase I:

$$UCL = \frac{(m-1)^2}{m}\beta_{\alpha,p/2,(m-p-1)/2}, \tag{2.31}$$

where β is the beta distribution with p/2 and (m − p − 1)/2 degree of freedom at significance level alpha (α).

Conversely at Phase II the limit is placed at

$$UCL = \frac{p(m+1)(m-1)}{m^2 - mp}F_{\alpha,p,m-p}. \tag{2.32}$$

Presumably, the traditional calculation of S is limited for the lack of subgroups, wherefore many estimators have been suggested.

Sullivan and Woodall (1996a) examined the use of the cumulative sum of differences regarding the mean by its transpose:

$$S_{sw} = \frac{\sum_{k=1}^{m} (x_i - \overline{x})(x_i - \overline{x})'}{m-1}. \tag{2.33}$$

On the other hand, Holmes and Mergen (1993) proposed the difference among consecutive observations instead of the difference respecting the mean:

$$S_{hm} = \frac{\begin{bmatrix} x_2 - x_1 \\ x_3 - x_2 \\ \vdots \\ x_m - x_{m-1} \end{bmatrix}\begin{bmatrix} x_2 - x_1 \\ x_3 - x_2 \\ \vdots \\ x_m - x_{m-1} \end{bmatrix}'}{2(m-1)}. \tag{2.34}$$

The following example shows the construction of the T^2 chart when n = 1.

Example 2.5

Bimetal thermostat has innumerable practical uses. These types of thermostats hold a bimetallic strip composed by two strips of different metals that convert the changing of temperature in mechanical displacement due to the difference in thermal expansion.

Certain type of strip composed of brass and steel is analyzed in a quality laboratory by testing the deflection, curvature, resistivity, and hardness in low and high expansion sides. Table 2.3 shows 28 samples taken by the quality control department.

The construction of the scatterplot matrices provides a graphical vision of the association of the variables (Fig. 2.12):

```
> pairs(bimetal1, labels = c("deflection","curvature","resistivity","Hardness low
  side","Hardness high side"))
```

The sample mean vector and the correlation matrix result in

$$\overline{X} = \begin{bmatrix} 21.02 \\ 40.02 \\ 15.19 \\ 22.02 \\ 26.01 \end{bmatrix}; \quad r = \begin{bmatrix} 1.00 & 0.61 & 0.38 & 0.40 & 0.60 \\ 0.61 & 1.00 & 0.59 & 0.51 & 0.85 \\ 0.38 & 0.59 & 1.00 & 0.22 & 0.50 \\ 0.40 & 0.51 & 0.22 & 1.00 & 0.32 \\ 0.60 & 0.85 & 0.50 & 0.32 & 1.00 \end{bmatrix}.$$

Table 2.3 Bimetal data of the Phase I

Sample	Deflection	Curvature	Resistivity	Hardness low expansion side	Hardness high expansion side	T^2 using "sw"	T^2 using "hm"
1	21.15	40.24	14.95	22.24	26.24	1.82	1.66
2	21.10	39.99	14.79	21.62	25.92	1.05	0.82
3	20.95	39.82	14.91	22.04	25.95	1.51	1.45
4	21.03	40.01	14.89	21.74	26.19	8.63	8.27
5	21.21	40.03	15.03	22.32	25.86	8.69	7.72
6	21.37	40.31	15.21	22.03	26.08	2.08	2.44
7	20.70	39.90	14.75	21.67	25.86	3.57	3.77
8	20.87	39.89	15.04	21.89	26.02	7.94	9.17
9	21.27	40.14	15.20	22.27	26.23	8.16	8.24
10	20.97	40.13	14.98	22.11	26.22	3.39	4.24
11	21.34	40.20	14.91	21.99	25.89	2.69	2.17
12	20.92	39.87	14.90	21.76	25.93	6.34	5.84
13	20.83	40.00	15.15	22.20	26.02	4.71	4.66
14	20.84	39.90	15.06	22.08	26.07	2.09	1.8
15	20.95	40.16	14.97	22.20	26.25	4.85	5.14
16	20.75	39.80	14.71	22.01	25.66	11.57	11.84
17	21.00	40.05	15.10	22.36	26.10	1.13	1.29
18	21.21	40.26	15.05	22.15	26.17	1.55	1.14
19	21.03	39.87	14.98	22.05	26.07	8.74	10.92
20	21.01	39.84	14.97	21.89	26.19	11.4	12.49
21	21.08	40.00	14.78	22.20	25.90	0.77	0.74
22	21.08	39.78	14.96	22.02	26.09	5.61	5.30
23	20.69	39.77	14.92	21.91	25.87	5.18	4.29
24	20.88	39.85	15.00	21.79	26.00	3.53	3.32
25	21.01	40.02	15.06	21.92	26.08	9.24	7.10
26	21.01	39.95	14.78	22.02	25.86	5.07	4.34
27	21.07	40.08	15.40	22.15	26.06	1.55	1.43
28	20.97	39.87	14.99	21.77	25.91	2.14	2.12

The computation of S_{sw} is as follows:

$$S_{sw} = \frac{\sum_{k=1}^{m}(x_i-\bar{x})(x_i-\bar{x})'}{m-1} = \frac{1}{28-1}\left\{\begin{bmatrix}20.84\\39.84\\14.98\\21.88\\25.87\end{bmatrix}' - \begin{bmatrix}21.02\\40.04\\15.19\\22.02\\26.01\end{bmatrix}'\right\} \times \left\{\begin{bmatrix}20.84\\39.84\\14.98\\21.88\\25.87\end{bmatrix} - \begin{bmatrix}21.02\\40.04\\15.19\\22.02\\26.01\end{bmatrix}\right\} +$$

$$\ldots + \left\{\begin{bmatrix}21.14\\39.93\\15.19\\22.02\\26.01\end{bmatrix}' - \begin{bmatrix}21.02\\40.04\\15.19\\22.02\\26.01\end{bmatrix}'\right\} \times \left\{\begin{bmatrix}21.14\\39.93\\15.19\\22.02\\26.01\end{bmatrix} - \begin{bmatrix}21.02\\40.04\\15.19\\22.02\\26.01\end{bmatrix}\right\}.$$

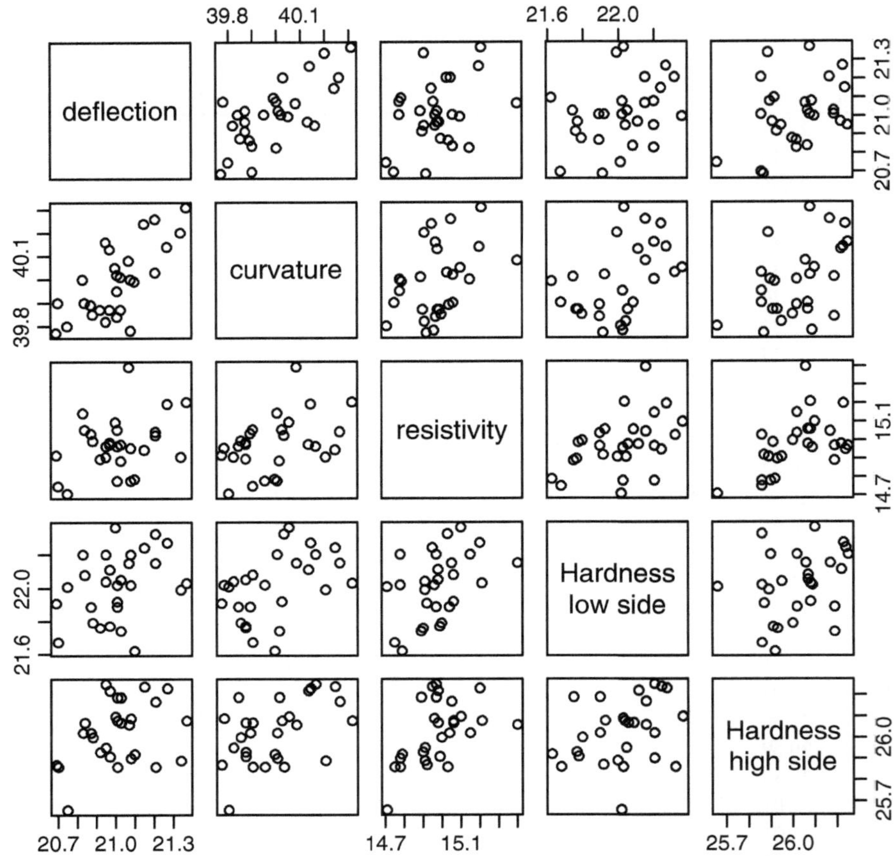

Fig. 2.12 Scatterplot matrices of the bimetal1 dataset

Finally

$$S_{sw} = \begin{bmatrix} 0.092 & 0.025 & 0.038 & 0.028 & 0.027 \\ 0.026 & 0.019 & 0.026 & 0.016 & 0.017 \\ 0.038 & 0.026 & 0.106 & 0.016 & 0.023 \\ 0.028 & 0.016 & 0.016 & 0.054 & 0.011 \\ 0.027 & 0.017 & 0.023 & 0.011 & 0.022 \end{bmatrix}.$$

So, the T^2 statistics is calculated as

$$T^2 = (X - \bar{X})'(S)^{-1}(X - \bar{X}),$$

$$T_1^2 = \left\{ \begin{bmatrix} 20.84 \\ 39.84 \\ 14.98 \\ 21.88 \\ 25.87 \end{bmatrix}' - \begin{bmatrix} 21.02 \\ 40.04 \\ 15.19 \\ 22.02 \\ 26.01 \end{bmatrix}' \right\} \times \begin{bmatrix} 0.092 & 0.025 & 0.038 & 0.028 & 0.027 \\ 0.026 & 0.019 & 0.026 & 0.016 & 0.017 \\ 0.038 & 0.026 & 0.106 & 0.016 & 0.023 \\ 0.028 & 0.016 & 0.016 & 0.054 & 0.011 \\ 0.027 & 0.017 & 0.023 & 0.011 & 0.022 \end{bmatrix}^{-1}$$
$$\times \left\{ \begin{bmatrix} 20.84 \\ 39.84 \\ 14.98 \\ 21.88 \\ 25.87 \end{bmatrix} - \begin{bmatrix} 21.02 \\ 40.04 \\ 15.19 \\ 22.02 \\ 26.01 \end{bmatrix} \right\},$$

$T_1^2 = 1.82$, and so forth for the others (that can be found in Table 2.3). On the other hand, to calculate S_{hm}

$$S_{sw} = \frac{\begin{bmatrix} x_2 - x_1 \\ x_3 - x_2 \\ \vdots \\ x_{m-1} - x_{m-2} \end{bmatrix} \begin{bmatrix} x_2 - x_1 \\ x_3 - x_2 \\ \vdots \\ x_{m-1} - x_{m-2} \end{bmatrix}'}{2(m-1)} = \frac{1}{2(28-1)}$$
$$\begin{bmatrix} 20.89 - 20.84 & 39.94 - 39.84 & \cdots & 25.97 - 25.87 \\ 21.13 - 20.89 & 40.12 - 39.94 & \cdots & 26.11 - 25.97 \\ \vdots & \vdots & \vdots & \vdots \\ 21.14 - 20.96 & 39.93 - 40.03 & \ldots & 25.98 - 25.94 \end{bmatrix}$$
$$\times \begin{bmatrix} 20.89 - 20.84 & 39.94 - 39.84 & \cdots & 25.97 - 25.87 \\ 21.13 - 20.89 & 40.12 - 39.94 & \cdots & 26.11 - 25.97 \\ \vdots & \vdots & \vdots & \vdots \\ 21.14 - 20.96 & 39.93 - 40.03 & \ldots & 25.98 - 25.94 \end{bmatrix}'$$

$$S_{hm} = \begin{bmatrix} 0.090 & 0.029 & 0.041 & 0.027 & 0.031 \\ 0.029 & 0.021 & 0.031 & 0.017 & 0.018 \\ 0.041 & 0.031 & 0.121 & 0.007 & 0.026 \\ 0.027 & 0.017 & 0.007 & 0.065 & 0.012 \\ 0.031 & 0.018 & 0.026 & 0.012 & 0.021 \end{bmatrix}.$$

```
[1] "Hotelling Control Chart"
$ucl
[1] 14.53
$t2
     [,1]
 [1,]  1.82
 [2,]  1.05
...
 [27,]  1.55
 [28,]  2.14
$Xmv
[1] 21.02 40.02 15.19 22.02 26.01
$covariance
      [,1]  [,2]  [,3]  [,4]  [,5]
[1,] 0.092 0.025 0.038 0.028 0.027
[2,] 0.025 0.019 0.026 0.016 0.017
[3,] 0.038 0.026 0.110 0.016 0.023
[4,] 0.028 0.016 0.016 0.054 0.011
[5,] 0.027 0.017 0.023 0.011 0.021
```

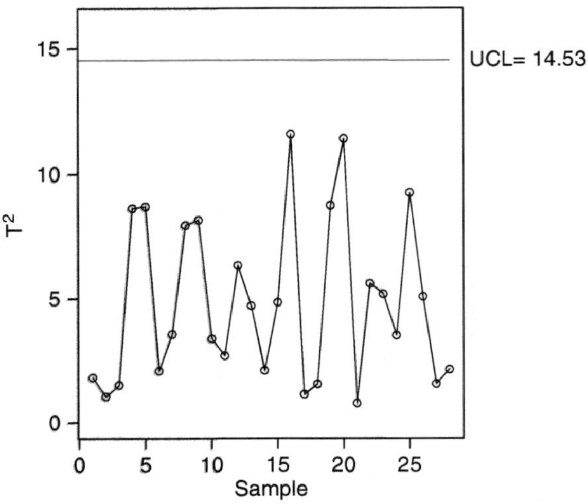

Fig. 2.13 Hotelling control chart with method = "sw" method and using the bimetal1 dataset

In the same manner to compute the statistics:

$$T_1^2 = \left\{ \begin{bmatrix} 20.84 \\ 39.84 \\ 14.98 \\ 21.88 \\ 25.87 \end{bmatrix}' - \begin{bmatrix} 21.02 \\ 40.04 \\ 15.19 \\ 22.02 \\ 26.01 \end{bmatrix}' \right\} \times \begin{bmatrix} 0.090 & 0.029 & 0.041 & 0.027 & 0.031 \\ 0.029 & 0.021 & 0.031 & 0.017 & 0.018 \\ 0.041 & 0.031 & 0.121 & 0.007 & 0.026 \\ 0.027 & 0.017 & 0.007 & 0.065 & 0.012 \\ 0.031 & 0.018 & 0.026 & 0.012 & 0.021 \end{bmatrix}^{-1}$$

$$\times \left\{ \begin{bmatrix} 20.84 \\ 39.84 \\ 14.98 \\ 21.88 \\ 25.87 \end{bmatrix} - \begin{bmatrix} 21.02 \\ 40.04 \\ 15.19 \\ 22.02 \\ 26.01 \end{bmatrix} \right\}$$

$T_1^2 = 1.66$ and so successively for the others.
With $\alpha = 0.05$ the UCL results in

$$UCL = \frac{(m-1)^2}{m} \beta_{\alpha,p/2,(m-p-1)/2} = \frac{(28-1)^2}{28} \beta_{0.05,5/2,(28-5-1)/2} = 14.53.$$

```
[1] "Hotelling Control Chart"
$ucl
[1] 14.53
$t2
      [,1]
 [1,] 1.66
 [2,] 0.82
...
[27,] 1.43
[28,] 2.12
$Xmv
[1] 21.02 40.02 15.19 22.02  26.01
$covariance
      [,1]  [,2]  [,3]  [,4]  [,5]
[1,] 0.090 0.029 0.041 0.027 0.030
[2,] 0.029 0.021 0.031 0.017 0.018
[3,] 0.041 0.031 0.120 0.007 0.026
[4,] 0.027 0.017 0.007 0.065 0.012
[5,] 0.030 0.018 0.026 0.012 0.021
```

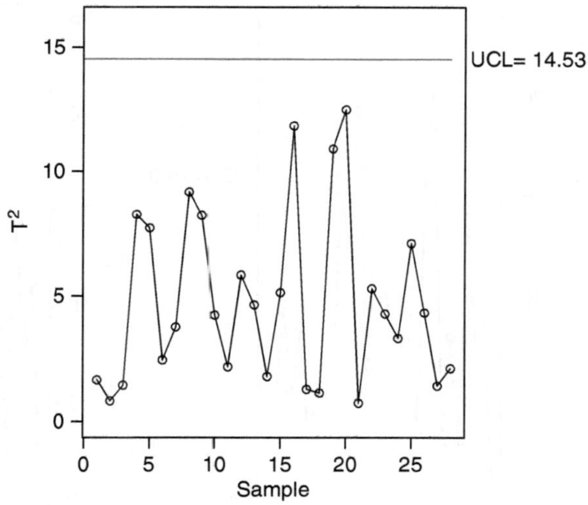

Fig. 2.14 Hotelling control chart with method = "hm" method and using the bimetal1 dataset

The mult.chart function detects automatically when x is a matrix or an array with depth n = 1 and computes S across any of the two methods that can be defined by the user using method = "sw" or "hm" or equally using the initials "s" or "h." Even if method is missing the default way is "sw":

```
> mult.chart(type = "t2", bimetal1, method = "sw", alpha = 0.05)
```

The output is shown in (Fig. 2.13)

In contrast to compute using the Holmes and Mergen (1993) method:

```
> mult.chart(type = "t2", bimetal1, method = "hm", alpha = 0.05)
```

obtaining (Fig. 2.14):

Notice that the function's output by using each method differs in the statistics and in the covariance matrix. In this example, comparing the two graphs, it can be seen that no significant difference was obtained from Holmes and Mergen (1993) and Sullivan and Woodall (1996a).

Now the extension of the example is possible by performing a control in the future production (Phase II) using the in-control mean and covariance obtained. The collected data of this production is stored in bimetal2.

Obviously it is needed to fix the in-control parameters:

```
> colm <- nrow(bimetal1)
>vec <- mult.chart(type = "t2", bimetal1, method = "sw", alpha = 0.05)$Xmv
```

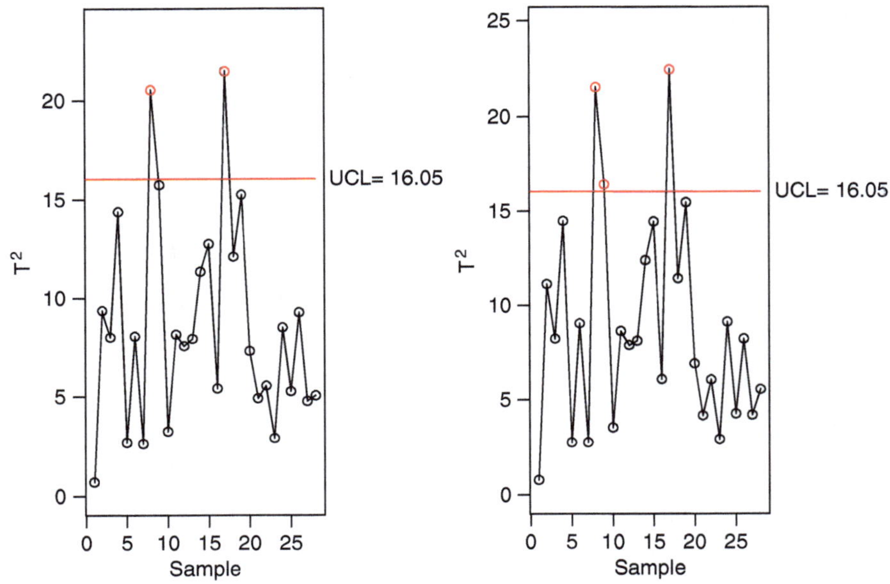

Fig. 2.15 Hotelling control chart in Phase II with both “sw” and “hm” method and using the bimetal2 dataset

First computing the covariance matrix according to the sw method:

>mat <− mult.chart(type = "t2", bimetal1, method = "sw", alpha = 0.05)
$covariance

and mat2 for the covariance with hm proposal.

>mat2 <− mult.chart(type = "t2", bimetal1, method = "hm", alpha = 0.05)
$covariance

> data("bimetal2")

To achieve both outputs in the same graphs:

par(mfrow = c(2,1))

> mult.chart(type = "t2", bimetal2, Xmv = vec, S = mat, method = "sw", alpha = 0.05)

> mult.chart(type = "t2", bimetal2, Xmv = vec, S = mat2, method = "hm", alpha = 0.05)

The chart using the sw method detects nonrandom shifts at the points 8 and 17 while that using the hm method detects the samples 8, 9, and 17.

Finally, both methods almost present similar sensitivity in this practical problem (Fig. 2.15).

2.7 Generalized Variance Control Chart

In the same manner as that in univariate control chart, the monitor of the process mean is coupled with a dispersion chart; monitoring the process variability results extremely useful in multivariate issues. This is because in the multivariate Shewhart chart it was assumed that the process dispersion remained constant. This hypothesis must be checked in practice.

To date various methods have been proposed for the simultaneous monitoring of variability but clearly the generalized variance chart is the most accepted. For more details see for example Alt (1985) or Montgomery (2004). The term *generalized variance* is known as the determinant of the covariance matrix.

This type of chart results by plotting the determinant of the covariance matrix along with the natural upper and lower control limits.

When the covariance matrix Σ is known the parameters of the chart result in

$$UCL = |\Sigma|\left(b_1 + 3{b_2}^{1/2}\right) \tag{2.35}$$

$$CL = b_1|\Sigma| \tag{2.36}$$

$$LCL = \max\begin{cases} |\Sigma|\left(b_1 - 3{b_2}^{1/2}\right) \\ 0 \end{cases}, \tag{2.37}$$

where

$$b_1 = \frac{1}{(n-1)^p}\prod_{j=1}^{p}(n-j) \tag{2.38}$$

and

$$b_2 = \frac{1}{(n-1)^{2p}}\prod_{j=1}^{p}(n-j)\left[\prod_{i=1}^{p}(n-i+2) - \prod_{i=1}^{p}(n-i)\right]. \tag{2.39}$$

Notice that n must be higher than the number of quality characteristics (p).

Frequently Σ is unknown and is estimated through S based on the relationship:

$$|S| = b_1|\Sigma| \tag{2.40}$$

Therefore the parameters result in

$$UCL = \frac{|S|}{b_1}\left(b_1 + 3{b_2}^{1/2}\right) \tag{2.41}$$

$$CL = |S| \tag{2.42}$$

$$LCL = \max\begin{cases} \frac{|S|}{b_1}\left(b_1 - 3{b_2}^{1/2}\right) \\ 0 \end{cases}. \tag{2.43}$$

Taking into account that S is positive-definite matrix, the LCL lacks of sense for negative values.

Example 2.6
Let us return to the carbon fiber data from Example 3.3 in which 30 samples of three quality characteristics of size n = 8 were taken.

In this case Σ is unknown and in consequence S was estimated:

$$S \times 100 = \begin{bmatrix} 0.25 & 0.36 & 0.67 \\ 0.36 & 1.40 & 1.00 \\ 0.67 & 1.00 & 5.92 \end{bmatrix}.$$

Then, the central line is $CL = |S| = 9.53 \times 10^{-7}$.
Secondly

$$b_1 = \frac{1}{(8-1)^3}\prod_{j=1}^{3}(8-j) = \frac{1}{343} \times 7 \times 6 \times 5 = 0.6122$$

$$b_2 = \frac{1}{(8-1)^6}\prod_{j=1}^{3}(8-j)\left[\prod_{i=1}^{3}(8-i+2) - \prod_{i=1}^{3}(8-i)\right] = \frac{1}{117649} \times 7 \times 6 \times 5$$
$$\times (9 \times 8 \times 7 - 7 \times 6 \times 5) = 0.5248$$

and finally

$$UCL = \frac{9.53 \times 10^{-7}}{0.6122}\left(0.6122 + 3 \times 0.5248^{1/2}\right) = 4.3386 \times 10^{-6}$$

$$LCL = \max\begin{cases} \frac{9.53 \times 10^{-7}}{0.6122}\left(0.6122 - 3 \times 0.5248^{1/2}\right) \\ 0 \end{cases}$$

$$LCL = \max\begin{cases} -2.4314 \times 10^{-6} \\ 0 \end{cases} = 0.$$

The elements of the sample covariance matrix and the corresponding determinant for each sample are presented in Table 2.4.

Table 2.4 Bimetal data for the generalized variance chart

	Means			Variances (×100)			Covariances (×100)			
Sample	Inner (X_1)	Thickness (X_2)	Length (X_3)	S_1	S_2	S_3	S_{12}	S_{13}	S_{23}	det(S)
1	1.03	1.08	50.16	0.15	1.19	2.98	−0.08	0.40	−0.42	3.10E−07
2	0.97	0.95	49.92	0.40	1.01	6.38	0.16	1.03	0.25	1.44E−06
3	1.01	1.05	50.14	0.17	0.99	1.96	0.31	0.38	0.81	7.00E−08
4	1.00	1.05	49.91	0.26	1.14	5.73	0.36	−0.29	0.82	5.00E−07
5	0.96	1.00	49.83	0.43	3.25	12.37	0.92	1.91	4.61	1.94E−06
6	1.03	1.07	50.05	0.30	1.52	1.69	0.43	0.30	−0.48	1.40E−07
7	0.96	1.02	49.95	0.17	0.58	4.34	0.03	0.34	0.74	2.70E−07
8	1.00	1.02	50.02	0.17	0.79	6.16	0.11	−0.11	0.32	6.90E−07
9	1.00	1.10	50.03	0.20	1.43	1.87	0.34	−0.07	−0.88	2.00E−07
10	0.99	1.02	50.00	0.13	0.53	6.58	0.11	0.18	−0.13	3.70E−07
11	1.01	1.10	50.01	0.18	1.31	3.41	0.11	0.19	0.36	7.20E−07
12	1.02	1.07	49.99	0.24	0.81	3.41	0.05	0.70	0.67	1.90E−07
13	0.97	1.00	49.96	0.48	2.36	17.72	0.98	2.44	5.17	8.60E−07
14	1.01	1.05	50.04	0.13	1.08	7.20	0.18	0.16	1.98	3.60E−07
15	1.00	1.06	50.02	0.24	1.14	7.80	0.26	1.01	0.43	6.70E−07
16	1.00	1.03	49.99	0.39	1.66	3.69	0.71	0.98	1.27	9.00E−08
17	1.00	1.04	49.99	0.10	1.27	7.71	0.00	0.44	−2.09	3.20E−07
18	0.98	1.00	49.94	0.18	1.56	5.40	0.26	0.85	2.22	1.30E−07
19	0.98	0.96	49.93	0.24	1.61	5.68	0.55	0.18	0.55	4.30E−07
20	1.01	1.07	50.02	0.37	2.55	4.91	0.64	1.16	3.33	3.00E−08
21	0.98	1.03	49.96	0.28	0.39	7.21	0.15	1.39	0.64	3.00E−08
22	0.99	1.04	50.07	0.23	2.46	8.24	0.60	0.70	1.74	1.20E−06
23	0.95	0.92	49.86	0.41	1.82	2.69	0.73	0.40	0.32	4.00E−07
24	1.00	1.09	50.05	0.15	0.75	9.27	0.12	0.69	−0.29	5.20E−07
25	0.99	1.01	49.96	0.51	1.87	7.08	0.56	1.56	1.63	1.50E−06
26	0.99	1.02	49.89	0.12	0.75	7.04	0.19	0.59	1.34	2.10E−07
27	0.99	1.03	49.84	0.24	3.80	7.47	0.72	0.87	2.20	1.61E−06
28	1.01	1.04	49.97	0.06	0.80	2.46	0.14	0.08	0.05	7.00E−08
29	1.03	1.10	50.07	0.19	1.29	2.38	0.43	0.36	0.72	1.20E−07
30	1.01	1.08	49.97	0.33	1.75	6.78	0.69	1.27	2.73	1.80E−07

The points to be plotted are the determinants of the covariance of each sample. For instance for the first sample:

$$\det(S_1) = \left\| \begin{bmatrix} 0.15 & -0.08 & 0.40 \\ -0.08 & 1.19 & -0.42 \\ 0.40 & -0.42 & 2.98 \end{bmatrix} \right\| = 3.10 \times 10^{-7}.$$

Performing in R is done through the gen.var function that only requires as argument an array of dimensions: $p \times m \times n$. For instance (Fig. 2.16):

```
> gen.var(carbon1)
```

Then R returns:

All points fall inside the control limits; therefore, there is no signal of out-of-control associated to the process variability.

```
[1] "Generalized Variance Control
Chart"
$`Upper Control Limit`
[1] 4.3e -06
$`Lower Control Limit`
[1] 0
$stat
     [,1]
 [1,] 3.1e -07
 [2,] 1.4e -06
...
[29,] 1.2e -07
[30,] 1.8e -07
```

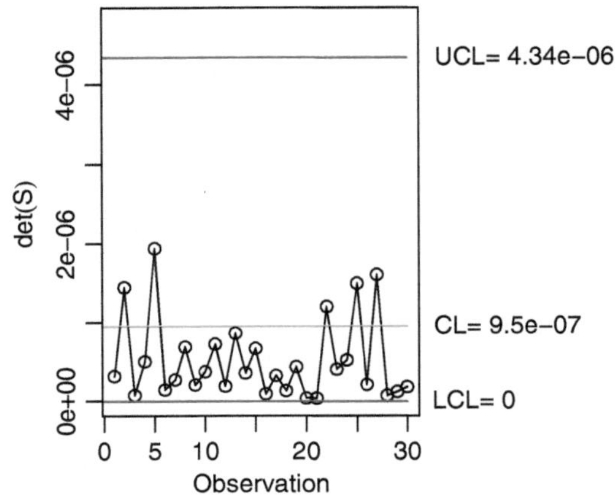

Fig. 2.16 Generalized variance control chart using the carbon1 dataset

2.8 Multivariate Exponentially Weighted Moving Average Control Chart

MEWMA is the natural multivariate extension of the EWMA chart proposed by Roberts (1959). It was introduced by Lowry et al. (1992) and is more sensible in detecting nonrandom changes in the process and based on the principle of the weighted average of the previously observed vectors.

Despite the fact that it is used mainly for individual observations (n = 1) it can be utilized in rational subgroup case as it will be explained later. It is also a chart for Phase II.

The MEWMA chart has the statistics:

$$T^2 = Z_i'\Sigma_{Z_i}^{-1}Z_i > h, \tag{2.44}$$

where

$$Z_i = \lambda X_i + (1 - \lambda)X_{i-1} \tag{2.45}$$

being $Z_o = 0$, λ is diagonal $p \times p$ matrix of the smoothing constant with $0 < \lambda_i \leq 1$, although in practice there is no reason to employ different values of λ in the same problem. Practically, the most often used value of λ is 0.1.

In a particular case, when rational subgroups are obtained, i.e., n > 1, just replace X_i by $\bar{X}_i$.

Lowry et al. (1992) provide two alternatives to compute the Σ_z, the exact covariance matrix:

$$\Sigma_{Z_i} = \frac{\lambda\left[1 - (1 - \lambda)^{2i}\right]}{2 - \lambda}(\Sigma) \tag{2.46}$$

and the named asymptotic covariance matrix

$$\Sigma_{Z_i} = \frac{\lambda}{2 - \lambda}(\Sigma) \tag{2.47}$$

the first one having a better performing.

Moreover, they point out that the ARL performance of the chart depends only on noncentrality parameter θ:

$$\theta = \left[(\mu_1 - \mu_0)'\Sigma(\mu_1 - \mu_0)\right]^{1/2}, \tag{2.48}$$

where μ_1 is the mean vector for Phase II. Notice that when $\lambda = 1$ MEWMA chart is transformed on T^2 chart.

One of the main troubles on this chart is the selection of the h or UCL. Prabhu and Runger (1997) presented computed tables, based on the Markov chain approach, to choose the UCL according to the parameters λ, p, θ, and ARL.

On the other hand, Bodden and Rigdon (1999) proposed a FORTRAN program to compute either the UCL for given values of ARL, λ, and p or ARL for values of UCL, λ, and p. These programs can be obtained on StatLib site at http://lib.stat.cmu.edu/jqt/31-1.

Example 2.7
To illustrate the MEWMA chart, return to Example 3.3 of the carbon fiber tubes.

With

$$S \times 100 = \begin{bmatrix} 0.25 & 0.36 & 0.67 \\ 0.36 & 1.40 & 1.00 \\ 0.67 & 1.00 & 5.92 \end{bmatrix} \text{ it is easy to obtain}$$

$$S_{Z_1} \times 10^5 = \frac{0.1\left[1 - (1 - 0.1)^{2\times 1}\right]}{2 - 0.1} \frac{\begin{bmatrix} 0.25 & 0.36 & 0.67 \\ 0.36 & 1.40 & 1.00 \\ 0.67 & 1.00 & 5.92 \end{bmatrix}}{100}$$

$$= \begin{bmatrix} 2.49 & 3.59 & 6.69 \\ 3.59 & 14.49 & 10.2 \\ 6.69 & 10.25 & 9.21 \end{bmatrix}$$

$Z_1 = \lambda\overline{X}_1 + (1-\lambda)\overline{X}_{1-1} = 0.1\overline{X}_1 + (1-0.1)\overline{X}_{1-1}$ being
$\overline{X}_1 = [\,1.03 \quad 1.08 \quad 50.16\,] - [\,0.99 \quad 1.04 \quad 49.98\,]$ then:

$$Z_1 = \begin{bmatrix} 0.0034 \\ 0.0043 \\ 0.0172 \end{bmatrix} \text{ and finally}$$

$$T_1^2 = Z_1{}'\Sigma_{Z_1}^{-1}Z_1 = \begin{bmatrix} 0.0034 \\ 0.0043 \\ 0.0172 \end{bmatrix}' \times \left\{ \begin{bmatrix} 2.49 & 3.59 & 6.69 \\ 3.59 & 14.49 & 10.2 \\ 6.69 & 10.25 & 9.21 \end{bmatrix} /10^5 \right\}^{-1} \times \begin{bmatrix} 0.0034 \\ 0.0043 \\ 0.0172 \end{bmatrix}$$

$$= 0.6236$$

and so forth for all values of i.

Using the program by Bodden and Rigdon (1999) with ARL = 200, $\lambda = 0.1$, and p = 3, UCL = 10.81 is obtained.

The execution in R of the MEWMA control chart is furthermore through the mult.chart function specifying type = "mewma."

Another argument to be entered is lambda and in its absence the function works with the default value 0.1.

In the MEWMA chart the covariance matrix could be used in three different ways to estimate S in the same way as the T^2 is computed through the matrix of the mean sample covariance for rational subgroups, and for individual observations, using the methods by Sullivan and Woodall (1996b) and Holmes and Mergen (1993).

For the computation of the UCL, mult.chart uses the method suggested by Bodden and Rigdon (1999). A drawback is that the amount of the choices to select lambda, p, and ARL is limited as follows:

p for values 2,3,. . .,10
lambda for 0.1, 0.2,. . .,0.9
ARL only 200

However the user can enter as argument the desired UCL obtained for instance by Prabhu and Runger (1997) or Bodden and Rigdon (1999).

To carry out the previous example in R, just:

```
> mult.chart(type = "mewma", carbon1)
```

Then R prompts:

Notice in Fig. 2.17 that no alarm is given since no point falls beyond the UCL.

The assumption of the central limits theorem is not satisfied in case of individual observations; therefore in practice the normality assessing must be done.

Borror et al. (1999) proved how EWMA chart is robust regardless whether data follows a normal distribution or not. Later, Testik and Runger (2003) prove through a Monte Carlo simulation how the MEWMA chart is robust to non-normal data. That is, MEWMA is a nonparametric chart, so it can be used with suitable performance independently of the distribution of the data, the latter being one of the most striking properties.

```
[1] "MEWMA Control Chart"
$ucl
[1] 10.81
$t2
   [,1]
 [1,] 0.62
 [2,] 0.30
...
[29,] 0.07
[30,] 0.12
$Xmv
[1]  0.99  1.04 49.98
$covariance
      [,1]   [,2]   [,3]
[1,] 0.0025 0.0036 0.0067
[2,] 0.0036 0.0140 0.0100
[3,] 0.0067 0.0100 0.0590
```

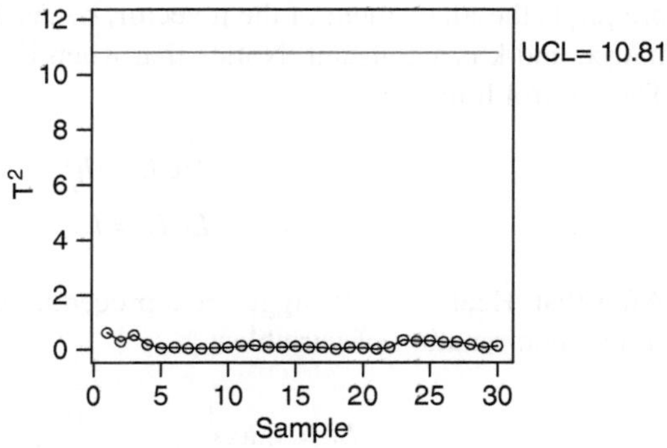

Fig. 2.17 MEWMA control chart with $\lambda = 0.1$ using the carbon1 dataset

2.9 Multivariate Cumulative Sum Control Chart

The MCUSUM control chart appears as the multivariate extension of the CUSUM control chart originally proposed by Page (1961). It is focused on improving the sensitivity regarding the previously introduced T^2 chart by detecting small shifts on the process and is based on the principle of accumulating information of the former observations. As well as the MEWMA chart, MCUSUM is a Phase II chart.

There are four main alternatives accepted to construct an MCUSUM chart which are exposed below.

The first of these suggestions was introduced by Woodall and Ncube (1985). They proposed the individual monitoring of the mean vector through the utilization of univariate CUSUM charts. Analogous to CUSUM there is also a two-side chart. Its statistics is given by

$$S_{i,j}^{-} = \min\left\{ \begin{array}{c} 0 \\ S_{i-1,j}^{-} + \dfrac{\bar{X}_{i,j} - \mu_{0,j}}{\sigma_{0,j}/\sqrt{n}} + k_j^{-} \end{array} \right\}$$
$$S_{i,j}^{+} = \max\left\{ \begin{array}{c} 0 \\ S_{i-1,j}^{+} + \dfrac{\bar{X}_{i,j} - \mu_{0,j}}{\sigma_{0,j}/\sqrt{n}} + k_j^{+} \end{array} \right\}, \qquad (2.49)$$

where $\mu_{0,j}$ is the jth element of the μ vector, $\sigma_{0,j}$ is the (j × j)th diagonal element of Σ matrix, and k is a constant. Notice that when i = 1 then $S_{i,j}^-$ and $S_{i,j}^+ = 0$.

The control limits are

$$\begin{aligned} UCL &= h_j^+ \\ LCL &= h_j^- . \end{aligned} \tag{2.50}$$

After that, Healy (1987) suggested a procedure to detect shifts in mean based on the linear combination of variables:

$$S_i = \max\left\{ \begin{array}{c} 0 \\ S_{i-1} + a' \ \overline{X}_i - k \end{array} \right\}, \tag{2.51}$$

where

$$a' = \frac{(\mu_1 - \mu_0)' \left(\frac{\Sigma_0}{n}\right)^{-1}}{\left[(\mu_1 - \mu_0)' \left(\frac{\Sigma_0}{n}\right)^{-1} (\mu_1 - \mu_0)\right]^{1/2}} \tag{2.52}$$

and

$$k = 0.5 \frac{(\mu_1 - \mu_0)' \left(\frac{\Sigma_0}{n}\right)^{-1} (\mu_1 - \mu_0)}{\left[(\mu_1 - \mu_0)' \left(\frac{\Sigma_0}{n}\right)^{-1} (\mu_1 - \mu_0)\right]^{1/2}} . \tag{2.53}$$

This chart includes the control limits:

$$UCL = h.$$

On the other hand, Crosier (1988) presented two multivariate procedures. Here we present the version of the better ARL performance.

The statistics is

$$T_i^2 = \left[S_i' \left(\frac{\Sigma}{n}\right)^{-1} S_i \right]^{1/2} > h, \tag{2.54}$$

where

$$S_i = \left\{ \begin{array}{cc} 0 & if \quad C_i \le k \\ (S_{i-1} + \bar{X}_i - \mu_o)\left(1 - \frac{k}{C_i}\right) & if \quad C_i > k \end{array} \right. , \tag{2.55}$$

where $S_0 = 0$, $k > 0$, and

$$C_i = \left[(S_{i-1} + \bar{X}_i - \mu_o)' \left(\frac{\Sigma}{n}\right)^{-1} (S_{i-1} + \bar{X}_i - \mu_o) \right]^{1/2} . \tag{2.56}$$

Likewise, the limit is

$$UCL = h.$$

Finally Pignatiello and Runger (1990) proposed likewise two MCUSUM charts, the following resulting as the better performance alternative:

$$T_i^2 = \max\left\{\begin{array}{c} 0 \\ \left[S_i'\left(\frac{\Sigma}{n}\right)^{-1} S_i\right]^{1/2} - kn_i \end{array}\right. \tag{2.57}$$

where

$$S_i = \sum_{j=i-n_i+1}^{i} \left(\overline{X}_i - \mu_0\right) \tag{2.58}$$

and

$$n_i = \left\{\begin{array}{ccc} n_{i-1} + 1 & if & T_{i-1}^2 > 0 \\ 1 & otherwise & \end{array}\right. \tag{2.59}$$

$$UCL = h.$$

Although we have introduced these four approaches, only the last two will be applied in this section.

Example 2.8
Returning to the example of the carbon data and beginning for Crosier (1988) method we have

$S_0 = 0$, $k > 0$, and

$$C_1 = \left[\begin{array}{l} \{[1.01\ \ 1.07\ \ 49.88] - [0.99\ \ 1.04\ \ 49.98]\}' \times \\ \left(\begin{bmatrix} 0.25 & 0.36 & 0.67 \\ 0.36 & 1.40 & 1.00 \\ 0.67 & 1.00 & 5.92 \end{bmatrix} /100/8\right)^{-1} \times \{[1.01\ \ 1.07\ \ 49.88] - [0.99\ \ 1.04\ \ 49.98]\} \end{array}\right]^{1/2}.$$
$$= 2.3591$$

After that, if $C_1 > k$ then

$$S_1 = \{[1.01\quad 1.07\quad 49.88] - [0.99\quad 1.04\quad 49.98]\}\left(1 - \frac{0.5}{2.3591}\right).$$
$$S_1 = [0.0138\quad 0.0266\quad -0.0828]$$

The statistics results in

$$T_1^2 = \left[\begin{array}{c} [0.0138 \quad 0.0266 \quad -0.0828]' \times \left(\begin{bmatrix} 0.25 & 0.36 & 0.67 \\ 0.36 & 1.40 & 1.00 \\ 0.67 & 1.00 & 5.92 \end{bmatrix} /100/8 \right)^{-1} \times \\ [0.0138 \quad 0.0266 \quad -0.0828] \end{array} \right]^{1/2}.$$

$$T_1^2 = 1.86$$

The other values are calculated in the same manner.

In the case of the Pignatiello and Runger (1990) MCUSUM we have $n_1 = 1$ and then $S_1 = \{[1.01 \quad 1.07 \quad 49.88] - [0.99 \quad 1.04 \quad 49.98]\}$, so

$$T_i^2 = \max \left\{ \begin{array}{c} 0 \\ \left[\begin{array}{c} [0.0138\ 0.0266\ -0.0828]' \times \left(\begin{bmatrix} 0.25\ 0.36\ 0.67 \\ 0.36\ 1.40\ 1.00 \\ 0.67\ 1.00\ 5.92 \end{bmatrix} /100/8 \right)^{-1} \\ \times [0.0138\ 0.0266\ -0.0828] \end{array} \right]^{1/2} - 0.5 \times 1. \end{array} \right.$$

$$T_i^2 = 1.86$$

The other values of T^2 can be computed in the same way.

The execution in R of the Crosier (1988) and Pignatiello and Runger (1990) MCUSUM charts it is also carried out using the mult.chart function specifying type = "mcusum" and "mcusum2," respectively.

Furthermore the arguments k and h must be entered although when these parameters are missing the function works with the default values 0.5 and 5.5, respectively. MCUSUM chart uses the same ways as T^2 and MEWMA to estimate the covariance matrix S.

In order to execute the previous example in R, just (Fig. 2.18):

```
> data("carbon2")
> Xmv <- mult.chart(carbon1, type = "t2") $Xmv
> S <- mult.chart(carbon1, type = "t2") $covariance
> mult.chart(type = "mcusum", carbon2, Xmv = Xmv, S = S)
```

Then R returns:

Specifying type = "mcusum2" R compute (Pignatiello and Runger 1990). The results obtained are presented in Fig. 2.19.

Finally signals of out-of-control are obtained; comparing the two results, it can be seen that Crosier (1988) chart provides a better sensitivity with signal since the seventh sample.

```
"MCUSUM Control Chart by
Crosier (1988)"
$ucl
[1] 5.5
$t2
     [,1]
 [1,] 1.86
 [2,] 2.54
 ...
[24,] 8.69
[25,] 9.41
$Xmv
   0.99      1.04      49.98
$covariance
       [,1]   [,2]   [,3]
[1,] 0.0025 0.0036 0.0067
[2,] 0.0036 0.0140 0.0100
[3,] 0.0067 0.0100 0.0590
```

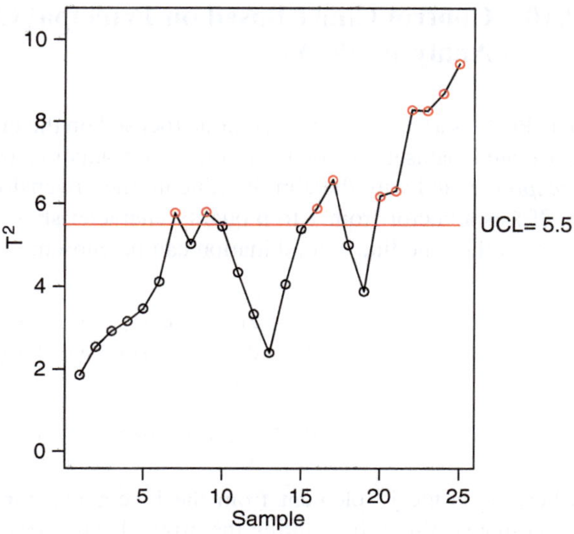

Fig. 2.18 MCUSUM control chart according to Crosier (1988) using the carbon1 dataset

```
"MCUSUM Control Chart by
Pignatiello (1990)"
$ucl
[1] 5.5
$t2
     [,1]
 [1,] 1.86
 [2,] 2.50
...
[24,] 7.16
[25,] 7.81
$Xmv
   0.99      1.04      49.98
$covariance
      [,1]   [,2]   [,3]
[1,] 0.0025 0.0036 0.0067
[2,] 0.0036 0.0140 0.0100
[3,] 0.0067 0. 0100 0.0590
```

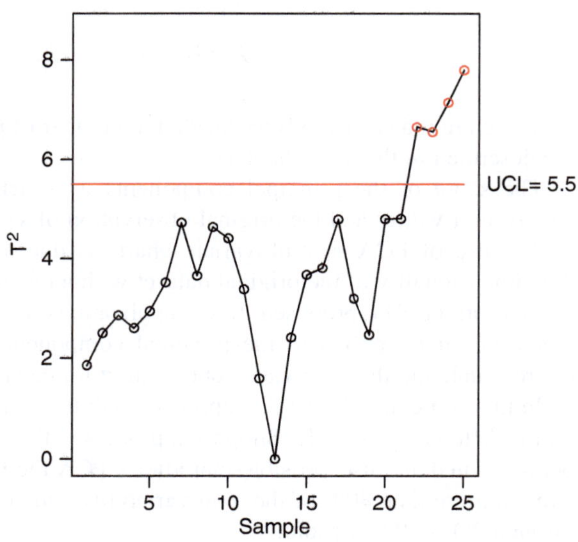

Fig. 2.19 MCUSUM control chart according to Pignatiello and Runger (1990) using the carbon1 dataset

2.10 Control Chart Based on Principal Component Analysis (PCA)

The PCA is a multivariate technique focused on the orthogonal transformation of a correlated dataset to obtain a linear combination of variables called *principal component* and with the aim of reducing the dimensionality.

If x is a vector from 1 to p quality characteristics with eigenvalues: $\lambda_1 \geq \lambda_2 \geq \ldots \geq \lambda_p$ then the linear combination can be chosen:

$$\begin{aligned} c_1 &= e_{11}x_1 + e_{12}x_2 + \cdots + e_{1p}x_p \\ c_2 &= e_{21}x_1 + e_{22}x_2 + \cdots + e_{2p}x_p \\ &\vdots \\ c_p &= e_{p1}x_1 + e_{p2}x_2 + \cdots + e_{pp}x_p \end{aligned}, \quad (2.60)$$

where e_{ij} is the jth element from the ith eigenvector and c_j the axes of the new coordinates system by rotating the original. This new axes represent the directions of maximum variability.

The principal components are chosen by maximizing the variance as much as possible.

The variance of the principal components is given by their eigenvalue and proportion of the variance explained is determined as

$$\lambda_j/(\lambda_1 + \lambda_2 + \ldots \lambda_p). \quad (2.61)$$

There are many methods to decide the number of principal components (which are described in the next chapter).

The score of the principal components c_j is determined by substituting the eigenvector values and the original observation of $x_1, x_2, \ldots, x_p$ in each c_j.

The use of PCA in multivariate charts is due to the feasibility of reducing the dimensionality of the original dataset without a significant loss of information.

Jackson (1991) proposed three applications of PCA in control chart: the Hotelling chart applied to the principal component scores, the control chart to the residual, and the univariate control chart for each score.

In this context, only the first approach will be studied. This application is based on the following principle. Suppose a process with 5 or 6 quality characteristics is being studied and it is possible that after a PCA the first two or three components explain more than 80% of the total variability and consequently can be controlled through 2D or 3D ellipsoid.

To illustrate this point the next example shows this clearly.

Example 2.9

Returning to the bimetal data introduced in Sect. 3.6.

To carry out the PCA in R just:

> eigen(covariance(bimetal1)) achieving the eigenvalues and eigenvectors
$values
[1] 0.170 0.066 0.040 0.015 0.002
$vectors

	[,1]	[,2]	[,3]	[,4]	[,5]
[1,]	0.597	0.548	0.511	−0.288	−0.004
[2,]	0.273	0.041	−0.074	0.525	−0.802
[3,]	0.641	−0.734	−0.075	−0.201	0.056
[4,]	0.299	0.395	−0.851	−0.140	0.108
[5,]	0.262	0.058	0.067	0.763	0.585

And to perform the summary of the principal components:

> summary(prcomp(bimetal1))

Then R shows:
Importance of components:

	PC1	PC2	PC3	PC4	PC5
Standard deviation	0.412	0.257	0.199	0.122	0.048
Proportion of variance	0.580	0.225	0.136	0.051	0.008
Cumulative proportion	0.581	0.806	0.942	0.992	1.000

This analysis can be complemented in a graphical way, for instance performing an elemental Pareto chart:
First get the variance through the standard deviation of the components:

> varian <− (prcomp(bimetal1)$sdev) ^ 2

Then, to store the proportion of variance and the cumulative proportion:

> perc <− varian / sum(varian)
> cumperc <− cumsum(perc)

Finally plotting the cumulative proportion as:

> plot(cumperc, type = "o", xlim = c(0.5, length(cumperc) + 0.5), ylim = c(0,1), xlab = "component", ylab = "percent") and adding the barplot
> barplot(perc, add = TRUE, width = 1, beside = TRUE, col = "gray", space = c(0,0.5))

As a result, the first two components are responsible for 80.61% of the variability. Therefore the original dimension of the problem has been reduced to a two-dimensional problem (Fig. 2.20).
Using

> prcomp(bimetal1)$x or
> predict(prcomp(bimetal1))

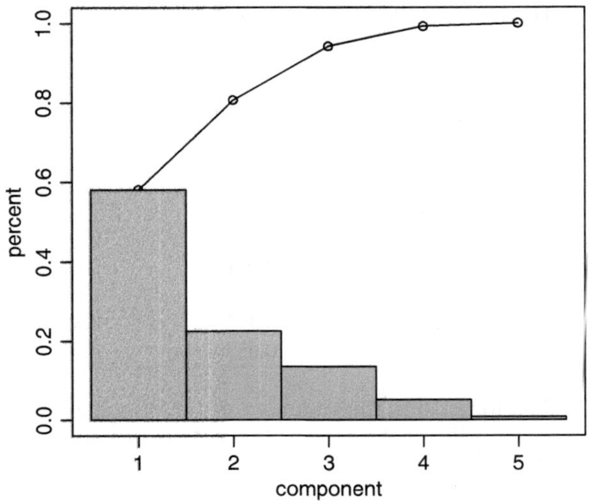

Fig. 2.20 Pareto chart of the principal components summary using the carbon1 dataset

Then R prompts the data from the principal components scores

	PC1	PC2	PC3	PC4	PC5
[1,]	−0.369	−0.013	0.052	−0.087	0.031
[2,]	−0.286	0.135	−0.046	0.020	0.022
...					
[27,]	−0.018	−0.101	−0.002	−0.039	−0.054
[28,]	−0.149	0.146	0.240	−0.037	0.018

Now, two alternatives can be taken:

1. Consider the parameters known or assume sufficiently a large dataset and execute a χ^2 control ellipse or a χ^2 chart.
2. Assume the parameters unknown and perform an F control ellipse or a T^2 control chart.

Suppose we decide to adopt the first one. To plot the first two components with the respective χ^2 confidence ellipse:

```
> a <− predict(prcomp(bimetal1))[,1:2]
> S <− covariance(a)
> Xmv <− colMeans(a)
```

Then plotting using the ellip function:

```
> plot(ellip(type="chi", a, alpha = 0.01),type="l", xlim = c(−1.6,1.6), ylim = c
  (−1,1), xlab= "z1", ylab= "z2")
> points(Xmv [1], Xmv [2], pch = 3) to include the centre or target
> points(a, cex = 0.75) and adding the points to the ellipse.
```

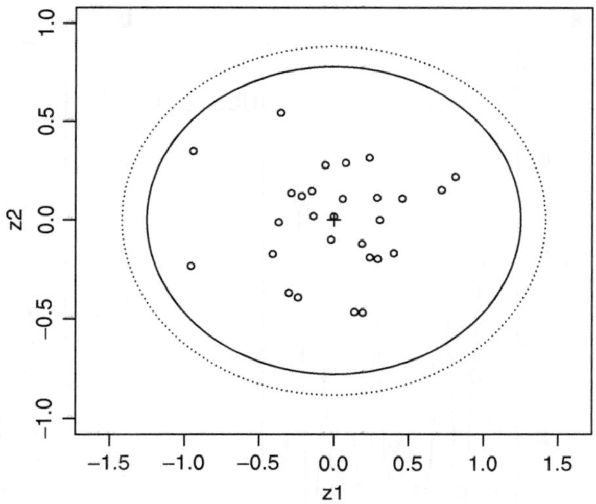

Fig. 2.21 Scatterplot for the principal component scores with the confidence ellipses in Phase I

On the contrary, if we select the second alternative which supposes the parameters of the distribution are unknown, then add an ellipsoid in dashed line to the existing one are unknown, then a wider ellipsoid results, as shown by the dashed line (Fig. 2.21):

```
points(ellip(type = "t2", a, alpha = 0.01), type = "l", lty = 3)
```

The control ellipsoid for the alternative with unknown parameters is less restrictive. In both cases, all the points are inside the confidence ellipsoid. Similar result can be obtained executing a χ^2 and Hotelling chart as can be seen in next figure (Fig. 2.22):

```
> par(mfrow = c(1,2))
> mult.chart(a, type = "chi", alpha = 0.01)
> mult.chart(a, type = "t2", alpha = 0.01)
```

Now analyzing the future production (Phase II) stored bimetal2 dataset, we have:

First, we use in the R graphics device the graph obtained in (Fig. 2.21) before the construction of the X^2 and Hotelling chart. Then to save the first two principal components scores:

```
> b <- cbind(predict(prcomp(bimetal2))[,1 : 2], 1 : nrow(bimetal2))
```

After that, to add the points to the existing graph:

```
> points(b[, 1], b[, 2], pch = 4, cex = 0.75)
```

And finally to incorporate the sample number:

```
> text(b[,1],b[,2], labels = b[,3], cex = 0.6, pos = 1, offset = 0.5)
```

unknown (Fig. 2.23):

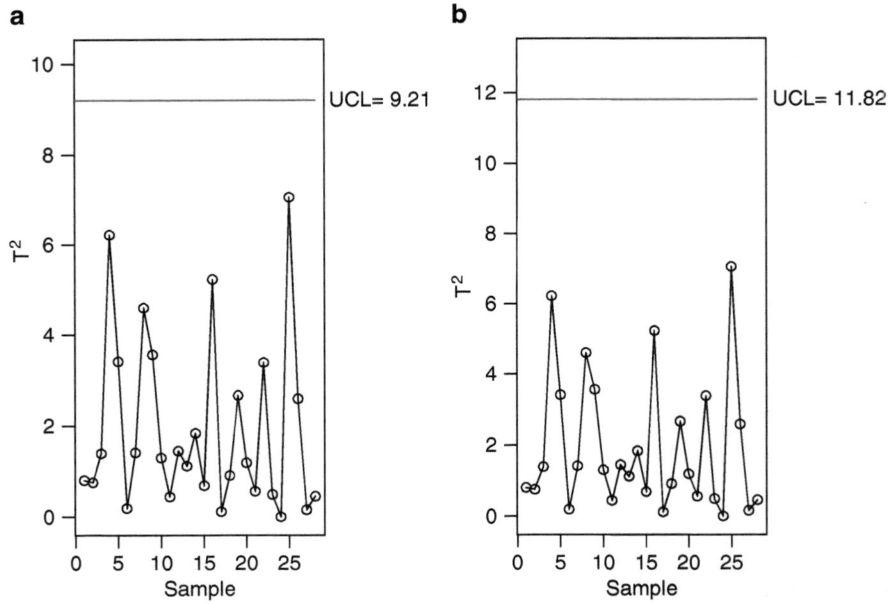

Fig. 2.22 (**a**) χ^2 and (**b**) Hotelling control chart of the principal component scores in Phase I

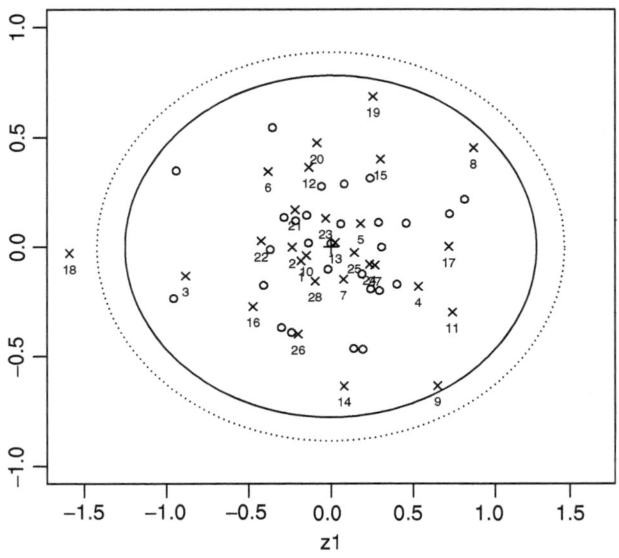

Fig. 2.23 Principal component scores with the confidence ellipses in Phase II

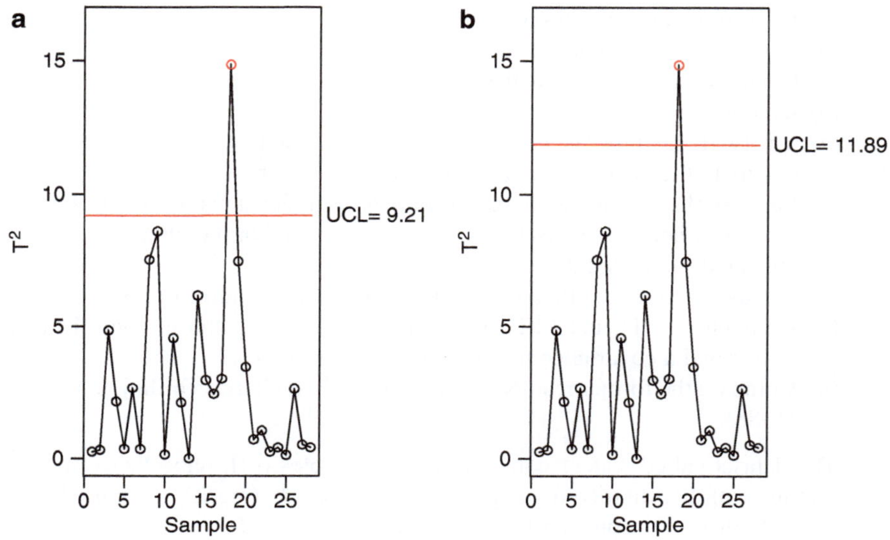

Fig. 2.24 (**a**) χ^2 and (**b**) Hotelling control chart of the principal component scores in Phase II

Likewise now using a Phase II χ^2 and Hotelling chart:

```
vec <- (mult.chart(a, type = "chi", alpha = 0.01)$Xmv)
mat <- (mult.chart(a, type = "chi", alpha = 0.01)$covariance)
par(mfrow = c(1,2))
mult.chart(b[,1:2], type = "chi", Xmv = vec, S = mat, alpha = 0.01)
mult.chart(b[,1:2], type = "t2", Xmv = vec, S = mat, alpha = 0.01)
```

As a result the 18th sample falls outside of ellipse contour and to the UCL. Notice that previous analysis with T^2 chart for individuals in Sect. 3.6 emitted an out-of-control signal in three moments.

In the output of the T^2 chart it can be shown that the source of the variability is associated to the first principal component (Fig. 2.24):

t2	decomp	ucl	p-value	1	2
[1,]	14.8626	7.9509	0.0006	1	0
[2,]	0.0123	7.9509	0.9125	2	0
[3,]	14.8749	11.8877	0.0000	1	2

2.11 Exercises

2.1. Two correlated quality characteristics are controlled in an industrial process. The indust1 and indust2 dataset represent the data obtained in two different moments. For indust1 dataset:

(a) Determine the correlation.
(b) Perform a scatterplot.
(c) Construct the 95th confidence ellipsoid.
(d) Compute a χ^2 chart.
(e) Verify if the process is under control using a Hotelling control chart. Compare the UCL with that achieved in one χ^2 chart.
Suppose that it was found that the process is under statistical control in the moment where the indust1 dataset was collected. For the future production collected in indust2 dataset:
(f) Use the confidence ellipsoid constructed in (c) to control in Phase II.
(g) Compute the T^2 and MEWMA control chart using the in-control mean vector and a covariance matrix obtained from indust1.
(h) Compare the former results with the MCUSUM chart according to Crosier (1988).

2.2. The dataset called water1 consists of five variables (pH, phosphates (mg/L), nitrates (mg/L), dissolved oxygen, and total solids (mg/L)) measured in a water quality test. Consider for all clauses alpha = 0.05.

(a) Determine the matrix of correlation coefficient.
(b) Construct a scatterplot matrix.
(c) Is it correct to use a χ^2 chart in this problem?
(d) Contruct a Hotelling control chart for water1 array. Is the process in statistical control?
(e) The former results achieved are carried out with the default method = " sw," for computing the covariance matrix in individual observation case. Compare this previous result with the ones obtained with the "hm" method.
(f) Are the MEWMA and MCUSUM capable to detect significantly causes in the process?
(g) The water2 represents a dataset composed by measures of a new stage. Construct the T^2 chart in Phase II. Is the process in control? Compare the UCL regarding the default alpha value 0.01.
(h) If any points fall beyond of the UCL, determine the source(s) of variation through the decomposition of T^2.
(i) Compute the MCUSUM by Crosier (1988) and according to Pignatiello and Runger (1990).
(j) Perform the MEWMA chart by using lambda 0.2 and 0.8.
(k) Perform a control chart based on PCA for the first two principal components. Compare the result with other charts. How many variance can be explained by the first two principal components?

2.3. Seven variables collected from a mechanical process are available at dataset named mech1 and mech2.

(a) Perform a control chart based on PCA analysis with alpha = 0.01. If any point falls outside the confidence ellipsoid identify it.

(b) How many variance can be explained by the first two principal components.
(c) Construct the Hotelling control chart using the hm method for computing the variance. If there is an out-of-control signal, determine the source.
(d) Is the MCUSUM more sensitive to the shift?

2.4. Three variables are measured with the aim to establish a multivariate monitoring program in a manufacturing process. The positive correlation between these quality characteristics was checked. The data collected are stored in the dataset called glass1.

(a) Construct the generalized variance chart. Discuss the results.
(b) Perform the T^2 control chart.
(c) Construct the MEWMA chart using lambda $= 0.2$ and 0.7.
(d) Perform the MCUSUM chart using UCL $= 5$.
(e) With the previous analysis accomplished: Is there evidence of out-of-control signal?

2.5. After a careful analysis performed in the previous exercise, the in-control state of the process was established with the aim of controlling future production.

(a) 32 samples were gathered in dataset glass2. Execute an analysis to determine if the process remains in statistical control.
(b) After that, 25 samples were collected. Perform the same analysis of the former clause. Compare the results.

Chapter 3
Multivariate Process Capability Indices (MPCI)

In the previous chapter it was explained how the evaluation of the process performance composed by many correlated quality characteristics should be carried out through a multivariate approach. In this chapter multivariate proposals of process capability are presented considering the most important developments in this field.

A capability index can be described as a ratio of the engineering specification to the process spread that provides information about satisfaction of the requirements.

Some of the earliest, significant works in this field were by Chan et al. (1991), Bothe (1992), and Pearn et al. (1992).

Since then, many indices have been proposed; among which the most recognized are by Hubele et al. (1991), Taam et al. (1993), Shahriari et al. (1995), and Chen (1994). Wang et al. (2000) performed a comparative study from these last methods and discussed their usefulness.

Wang and Chen (1998), Xekalaki and Perakis (2002), and Wang (2005) suggested indices based on principal component analysis as an extension of the univariate C_p, C_{pm}, C_{pk} and C_{pmk}, and Shinde and Khadse (2008) pointed out that the issue finding transformed the tolerance region for these indices.

Pearn and Kotz (2006) offered a review of this field and updated it in 2004 and Yum and Kim (2012) performed an extensive bibliographical review on process capability.

More recently, Pan and Lee (2010) proposed a modification to the Taam et al. (1993) index to avoid overestimation; Scagliarini (2011) studied the presence of measurement errors in indices base on PCA, and Tano and Vännman (2011) performed a comparison of the confidence intervals.

The number of approaches or proposals have increased significantly in recent years. Shinde and Khadse (2008) classified the indices into four categories based on:

1. Ratio of the volume tolerance region to a process region, e.g., Taam et al. (1993), Shahriari et al. (1995), Pan and Lee (2010), etc.

E. Santos-Fernández, *Multivariate Statistical Quality Control Using R*, SpringerBriefs in Statistics 14, DOI 10.1007/978-1-4614-5453-3_3,

2. The use of principal component analysis (PCA), e.g., Wang and Chen (1998), Xekalaki and Perakis (2002), Wang (2005), etc.
3. The probability of the nonconforming product as in Wierda (1993), Chen (1994), Chen et al. (2003), Castagliola and Castellanos (2005), etc.
4. Others.

3.1 The mpci Function

The measure of the process capability in the multivariate perspective can be implemented with the mpci function, which is a general function. This function allows to compute the most accepted capability indices as:

- Shahriari et al. (1995) vector
- Taam et al. (1993) index
- Pan and Lee (2010) proposal
- Wang and Chen (1998) indices employing PCA
- Xekalaki and Perakis (2002) indices
- Wang (2005) indices

The selection of the kind of index to use is done by specifying the argument index = "shah", "taam", "pan", "wang", "xeke" or "wangw" in the same order as they are introduced.

The function contains three compulsory arguments: x, which must be a matrix or data frame and the lower and upper specification limits, LSL, and USL, respectively. The target of the process could be specified and in case of missing values it is calculated as midpoint of the engineering tolerances.

In bivariate cases the logical argument graphic allows to achieve a graphical representation of the indices. For the specific case of p = 3 quality characteristics, the use of three-dimensional graph using the rgl package is illustrated in Sect. 3.5.

For the first three indices, alpha is the proportion of nonconforming products (conventionally fixed in 0.0027). In the case of the indices based on PCA, alpha is the significance level for the methods described below.

For these last indices the npc argument allows to specify the number of components to retain. The function is also capable of developing five methods to select the components by introducing *method = 1, 2, . . . , or 5* or the name of the routine, e.g., *method = "Percentage."*

After the execution mpci returns a list that contains a vector for the Shahriari et al. (1995) proposal, a list of four indices in indices that employ PCA and a single value for the Taam et al. (1993) and Pan and Lee (2010) indices.

The help of the function offers more details. See help (package = "MPCI") and for other examples see Santos-Fernández and Scagliarini (2012).

3.2 Multivariate Process Capability Vector

The Multivariate Process Capability Vector was introduced by Shahriari et al. (1995) based on the pioneer work of Hubele et al. (1991). It consists of a three-component vector which is defined as:

$$[CpM, PV, LI], \tag{3.1}$$

and is based on the assumption that the process follows a multivariate normal distribution.

The first component of this vector is CpM that is a ratio of the areas or volumes between the engineering tolerances and the modified process region.

$$CpM = \left[\frac{\prod_{i=1}^{p}(USL_i - LSL_i)}{\prod_{i=1}^{p}(UPL_i - LPL_i)}\right]^{1/p} \tag{3.2}$$

p being the number of quality characteristics.

Both areas and volumes are rectangles in bivariate process and rectangle prism in a three-dimensional case.

The area defined by the engineering tolerance is shown in Fig. 3.1 as the external rectangle. On the other hand, the modified process region is constructed as the smallest rectangle that circumscribes the ellipsoid or contour named *process region*. The ellipsoid is a probability density contour centered at the process mean, which is constructed by the spectral decomposition of the covariance matrix centered at the mean vector as it was shown in the previous chapter.

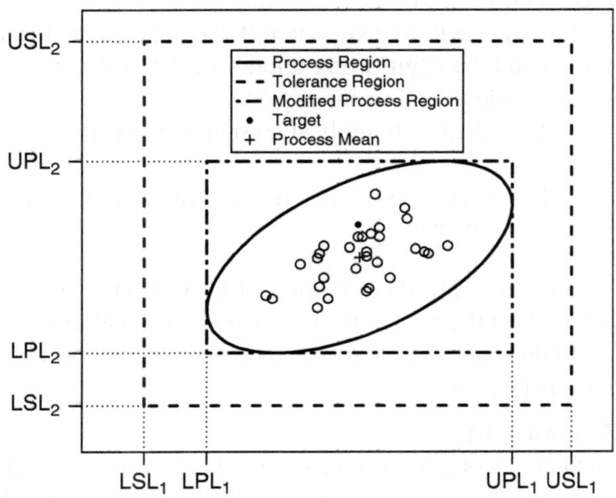

Fig. 3.1 Graphical representation of the modified process region

The borders of the process region: the lower process limits (LPL_i) and the upper process limits (UPL_i) are computed by solving the system of equation of the first derivatives of the quadratic form according to Nickerson (1994).

$$(X-\mu)'(\Sigma)^{-1}(X-\mu)=\chi^2_{\alpha,p} \tag{3.3}$$

with a χ^2 distribution with p degrees of freedom and significance level α.

The solutions of the equation for each dimension are given by:

$$LPL_i=\mu_i+\sqrt{\frac{\chi^2_{\alpha,p}\det\left(\Sigma_i^{-1}\right)}{\det\left(\Sigma^{-1}\right)}};\quad UPL_i=\mu_i+\sqrt{\frac{\chi^2_{\alpha,p}\det\left(\Sigma_i^{-1}\right)}{\det\left(\Sigma^{-1}\right)}} \tag{3.4}$$

where det() are the determinants and Σ_i^{-1} is the matrix achieved by deleting the ith column and row.

Values of CpM greater than 1, indicate that the modified process region is smaller than the engineering tolerance region.

The second component (PV) of the vector is the nearness between the target and the process mean, expressed by the hypothesis that

$$PV=P\left(T^2>\frac{p(m-1)}{m-p}F_{p,m-p}\right), \tag{3.5}$$

where

$$T^2=n(\overline{X}-\mu)'(S)^{-1}(\overline{X}-\mu) \tag{3.6}$$

and $F_{p,m-p}$ the F distribution with and $m-p$ degrees of freedom respectively.

PV takes values between 0 and 1, and values near zero point out that the process mean is distant to the process target.

Finally, the third component (LI) compares the locations of the modified process region and the engineering tolerance, showing when any part of the process region falls outside the tolerance region.

Values of $LI=0$ imply at least in one direction the tolerance region is exceeded.

$$LI=\begin{cases}1 & \text{if modified process region is contained within the engineering tolerance region.}\\ 0 & \text{otherwise}\end{cases}$$

Summing up, the Shahriari et al. (1995) vector provides a comparison of the volumes of the region, the closeness of the centers and the extensions of the regions.

In this example the computation of the Shahriari et al. (1995) vector in bivariate case is presented.

Example 3.1

In Sect. 2.4 is introduced an example of a dowel manufacturing process in which 40 samples of the diameter and the length were taken. The process has the following tolerances: $\mathrm{LSL}_1 = 0.47$ and $\mathrm{USL}_1 = 0.53$ for the diameter and $\mathrm{LSL}_2 = 0.90$ and $\mathrm{USL}_2 = 1.10$ for the length.

The mean vector and the covariance matrix are respectively:

$$\bar{x}' = \begin{bmatrix} 0.5009 & 1.0018 \end{bmatrix} \quad \text{and} \quad S = \begin{bmatrix} 4.9087\mathrm{e}-05 & 8.5849\mathrm{e}-05 \\ 8.5849\mathrm{e}-05 & 4.1994\mathrm{e}-04 \end{bmatrix}$$

According to the information given in the problem the engineering tolerances result is as follows: $LSL' = \begin{bmatrix} 0.47 & 0.90 \end{bmatrix}$ and $USL' = \begin{bmatrix} 0.53 & 1.10 \end{bmatrix}$.

Therefore, the target of the process can be estimated as the midpoint of the tolerances: Target' $= \begin{bmatrix} 0.50 & 1.00 \end{bmatrix}$.

The modified process region result for i = 1

$$LPL_1 = \mu_1 + \sqrt{\frac{\chi^2_{\alpha,p} \det\left(\Sigma_1^{-1}\right)}{\det\left(\Sigma^{-1}\right)}}$$

$$= 0.5009 + \sqrt{\frac{\chi^2_{0.0027,2} \times \det\left(4.1994\mathrm{e}-04^{-1}\right)}{\det\left(\begin{bmatrix} 4.9087\mathrm{e}-05 & 8.5849\mathrm{e}-05 \\ 8.5849\mathrm{e}-05 & 4.1994\mathrm{e}-04 \end{bmatrix}^{-1}\right)}} = 0.4767$$

The other values are obtained in the same fashion:

$$LPL_2 = 0.9313; \quad UPL_1 = 0.5250 \quad \text{and} \quad UPL_2 = 1.0723.$$

The area obtained from plotting these points is the minimum bounding rectangle of the confidence ellipsoid so called *modified process region*.

Then the CpM result:

$$CpM = \left[\frac{\prod_{i=1}^{p} (USL_i - LSL_i)}{\prod_{i=1}^{p} (UPL_i - LPL_i)}\right]^{1/p}$$

$$= \left[\frac{(0.53 - 0.47) \times (1.10 - 0.90)}{(0.5250 - 0.4768) \times (1.0723 - 0.9313)}\right]^2 = 1.3291$$

The second component of the vector (PV) is computed as:

$$PV = P(T^2 > \frac{p(m-1)}{m-p} F_{p,m-p}) = P(T^2 > \frac{2(40-1)}{40-2} F_{2,40-2})$$

$$= P(0.0159 > \frac{2(39)}{38} F_{2,38}) = 0.7351$$

since:

$$
\begin{aligned}
T^2 &= n(\overline{X}-\mu)'(S)^{-1}(\overline{X}-\mu) \\
&= \left(\begin{bmatrix}0.5009\\1.0018\end{bmatrix}-\begin{bmatrix}0.50\\1.00\end{bmatrix}\right)' \times \begin{bmatrix}4.9087e-05 & 8.5849e-05\\8.5849e-05 & 4.1994e-04\end{bmatrix}^{-1} \\
&\quad \times \left(\begin{bmatrix}0.5009\\1.0018\end{bmatrix}-\begin{bmatrix}0.50\\1.00\end{bmatrix}\right) \\
&= 0.0159
\end{aligned}
$$

In the calculation of the third component (LI) of the vector, the process specifications are compared to the tolerances are compared, being one if it satisfies the condition:

$$LSL_i < LPL_i \quad \text{and} \quad USL_i > UPL_i$$

In this case

$$0.47 < 0.4768; \quad 0.90 < 0.9313 \quad \text{and} \quad 0.53 > 0.5250; \quad 1.10 > 1.0723.$$

As per results the modified process region is contained by the tolerance region, therefore LI = 1. This last result can be verified in graphical form likewise.

Finally the Shahriari et al. (1995) vector results:

$$[CpM, PV, LI] = [\,1.3290 \quad 0.7351 \quad 1\,]$$

Since CpM is higher than 1, this indicates that the process modified region is smaller than the tolerance region. The value of PV is not enough near 0 to assert that there is a significant difference between the center of the process and the process target. Finally, the value of LI = 1 signifies that the process region is inside in the engineering region. Summarizing, the process was founded capable to fulfill the specifications.

The computation of this vector is done by using the function mpci and using the argument "shah" to specify the index to use:

```
> library("MPCI")
> data("dowel1")
> LSL <- c(0.47, 0.90); USL <- c(0.53, 1.10); Target <- c(0.50, 1.00)
```

Note that the tolerances are entered by introducing a vector of lower specifications and the other of upper specifications.

```
> mpci(index = "shah", dowel1, LSL, USL, Target, graph = TRUE)
```

The argument graph provides in a two-dimensional case (p = 2) a graphical representation. The output is shown below (Fig. 3.2).

The graph above shows the process control ellipse with its bounded rectangle and the engineering tolerance region.

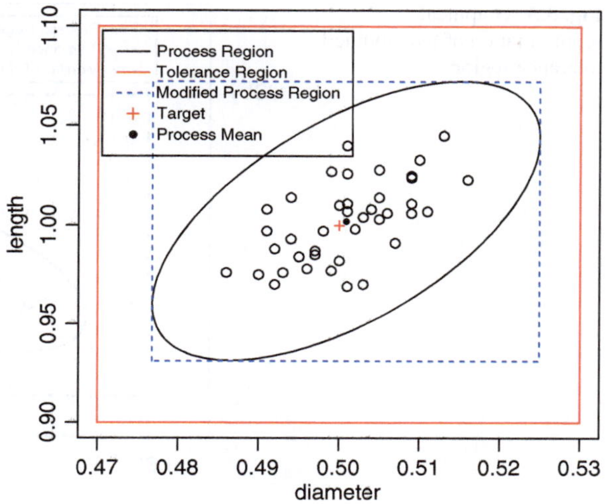

Fig. 3.2 Graphical representation of the areas in Shahriari's method
[1] "Shahriari et al. (1995) Multivariate Capability Vector"
$CpM
[1] 1.33
$PV
[,1]
[1,] 0.74
$LI
[1] 1

3.3 Multivariate Capability Index

Another widely accepted multivariate index is the MCpm proposed by Taam et al. (1993). It is defined as the ratio of the volumes of the ellipsoids of the modified tolerance region to the process region given by the control ellipsoid, see Fig. 3.3. On the contrary to the first component of the vector of Shahriari et al. (1995), which is computed as the ratio of the rectangles in bivariate case or hypercubes for more dimensions, the MCpm is the ratio of the ellipsoids.

The modified tolerance region is the largest ellipsoid constructed in the tolerance region and centered at the target.

The index is computed as:

$$MCpm = \frac{vol.(R_1)}{vol.(R_2)}, \tag{3.7}$$

where R_1 and R_2 are the modified tolerance region and the confidence ellipsoid respectively. This ratio can be estimated as:

$$MCpm = \frac{Cp}{D} \tag{3.8}$$

with

$$Cp = \frac{vol.(tolerance\,region)}{vol.(estimated\,process\,region)} \tag{3.9}$$

and the numerator is the hyperellipsoid with volume determined as:

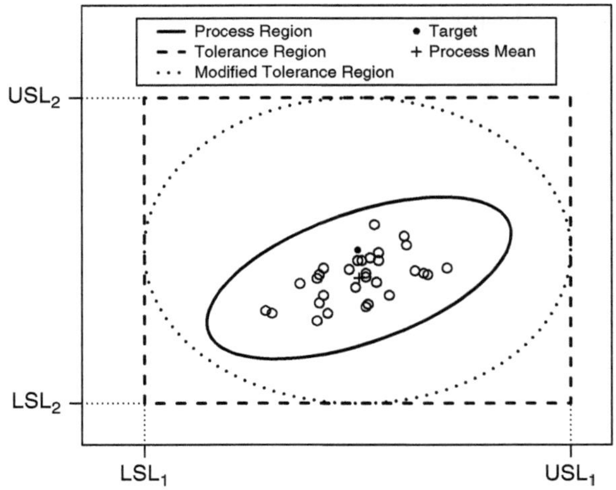

Fig. 3.3 Graphical representation of the modified tolerance region

$$vol.(tolerance\,region) = \frac{2\pi^{p/2}\prod_{j=1}^{p} l_j}{p\Gamma(p/2)} \tag{3.10}$$

where l_j is the length of the semi-axes. On the other hand:

$$vol.(estimated\,process\,region) = |S|^{1/2}(\pi K)^{p/2}[\Gamma(p/2+1)]^{-1}, \tag{3.11}$$

where K is the percentile of the χ^2 distribution and

$$D = \left[1 + \frac{m}{m-1}(\overline{X}-\mu)'(S)^{-1}(\overline{X}-\mu)\right]^{1/2}. \tag{3.12}$$

Therefore,

$$MCpm = \frac{vol.(R_1)}{\left\{|S|^{1/2}(\pi K)^{p/2}[\Gamma(p/2+1)]^{-1}\right\} * \left[1+\frac{m}{m-1}(\overline{X}-\mu)'(S)^{-1}(\overline{X}-\mu)\right]^{1/2}} \tag{3.13}$$

When the value of the index is >1 and the process mean vector is equal to the target, this implies that the process volume is smaller than the modified tolerance region.

Example 3.2

To illustrate the computation of the MCpm index, recall the dowel data analyzed in the previous section.

$$vol.(R_1) = \prod_{i=1}^{p} (USL_i - LSL_i) \times \frac{(2\pi)^{p/2}}{p \times \Gamma(p/2)}$$
$$= (0.53 - 0.47) \times (1.10 - 0.90) \times \frac{(2 \times 3.1416)^{2/2}}{2 \times \Gamma(2/2)} = 0.0094$$

then

$$|S|^{1/2}(\pi K)^{p/2}[\Gamma(p/2+1)]^{-1} = \left\{ \det\left(\begin{bmatrix} 4.9087\text{e}-05 & 8.5849\text{e}-05 \\ 8.5849\text{e}-05 & 4.1994\text{e}-04 \end{bmatrix}^{-1} \right) \right\}^{1/2}$$
$$\times (3.1416 \times \chi_{\alpha,p})^{2/2}[\Gamma(2/2+1)]^{-1} = 0.0043$$

and

$$\left[1 + \frac{m}{m-1}(\overline{X} - \mu)'(S)^{-1}(\overline{X} - \mu)\right]^{1/2}$$
$$= \left\{ 1 + \frac{40}{40-1} \left(\begin{bmatrix} 0.5009 \\ 1.0018 \end{bmatrix} - \begin{bmatrix} 0.50 \\ 1.00 \end{bmatrix} \right)' \times \begin{bmatrix} 4.9087\text{e}-05 & 8.5849\text{e}-05 \\ 8.5849\text{e}-05 & 4.1994\text{e}-04 \end{bmatrix}^{-1} \right.$$
$$\left. \times \left(\begin{bmatrix} 0.5009 \\ 1.0018 \end{bmatrix} - \begin{bmatrix} 0.50 \\ 1.00 \end{bmatrix} \right) \right\}^{1/2} = 1.0081$$

Then,

$$MCpm = \frac{0.0094}{0.0043 \times 1.0081} = 2.1860.$$

To perform the example in R, the argument index = "taam" must be specified.

> mpci(index = "taam", dowel1, LSL, USL, Target, graph = TRUE)

Then R prompts:

This index finds also capable the process and obtain a greater value than CpM (Fig. 3.4).

3.4 Revision of the Multivariate Capability Index

A more recent proposal is due to Pan and Lee (2010), which is a special case of the Taam et al. (1993) index. The authors pointed out that the Taam et al. (1993) index could suffer an overestimation if the quality characteristics are not independent. In this case the tolerance region is given by:

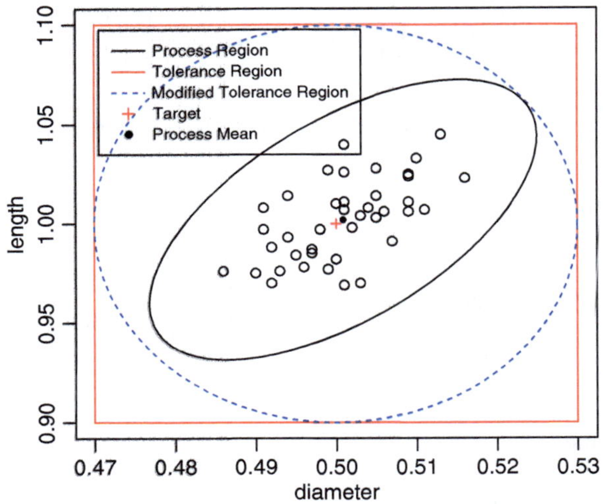

Fig. 3.4 Graphical representation of the areas in Taam's method
[1] "Taam et al. (1993) Multivariate Capability Index (MCpm)"
$MCpm
[,1]
[1,] 2.19

$$(X - T)'(A^*)^{-1}(X - T) = \chi^2_{p,1-\alpha}, \tag{3.14}$$

where

$$A^*_{ij} = r_{ij}\left(\frac{USL_i - LSL_i}{2\sqrt{\chi^2_{p,1-\alpha}}}\right)\left(\frac{USL_j - LSL_j}{2\sqrt{\chi^2_{p,1-\alpha}}}\right) \tag{3.15}$$

and r_{ij} is the correlation coefficient between i and j. Finally the proposed index results in:

$$NMCpm = \left(\frac{|A^*|}{|S|}\right)^{1/2} \tag{3.16}$$

The figure below shows the slanted ellipsoid with longdash (lty = 5) as line type (Fig. 3.5).

Example 3.3
Recall the dowel1 dataset to present the index in bivariate case.
For i = 1 and j = 1

$$A^*_{11} = r_{11}\left(\frac{USL_1 - LSL_1}{2\sqrt{\chi^2_{p,1-\alpha}}}\right)\left(\frac{USL_1 - LSL_1}{2\sqrt{\chi^2_{2,0.9973}}}\right) = 1 \times \left(\frac{0.53 - 0.47}{2\sqrt{11.83}}\right)^2 = 7.6e^{-5},$$

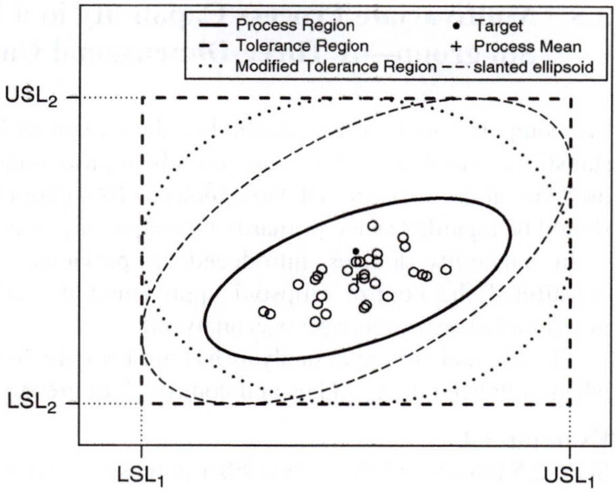

Fig. 3.5 Graphical representation of the revised region with the Taam's modified tolerance region

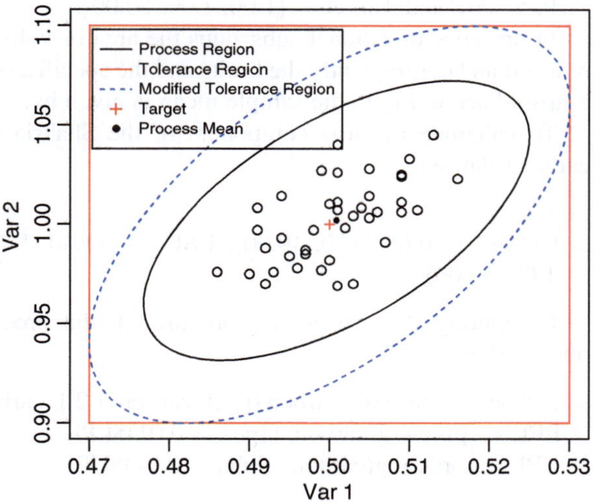

Fig. 3.6 Graphical representation of the areas in Pan's method
[1] "Pan and Lee (2010) Multivariate Capability Index (NMCpm)"
$NMCpm
[,1]
[1,] 1.75

and so on for the others.

Then,

$A^* = \begin{bmatrix} 7.6e-5 & 15.2e-5 \\ 15.2e-5 & 84.5e-5 \end{bmatrix}$ and det(S) $= 1.32\text{e}{-8}$. Finally $NMCpm = 1.77$.

The computation in R is as follows (Fig. 3.6):

```
> mpci(index = "pan", dowel1, LSL = LSL, USL = USL, graph = TRUE)
```

3.5 Multivariate Process Capability in a Presence of Rational Subgroup—A Three-Dimensional Case

The computation of process capability indices requires that the process operates under statistical control. Therefore sometimes both process capability and control chart are used simultaneously. In fact Montgomery (2004) pointed out that "the control chart should be regarded as the primarily technique of process capability analysis."

In capability indices introduced in previous sections the process region constituted the control ellipsoid approached in Sect. 3.4; but so far only the individual observation case was analyzed.

The rational subgroup analysis has not been diversified yet, but it can be useful when both capability studies and control chart are studied together.

Example 3.4

Sect. 2.5 introduced the carbon fiber process, in which 28 samples of three quality characteristics were collected. The sample size of each sample was eight. In this process the specifications are given by: LSL = [0.60, 0.30, 49.00], USL = [1.40, 1.70, 51.00], and Target = [1.00, 1.00, 50.00].

In presence of rational subgroups the area or volume of the swarm of points is reduced and consequently the limits and the specifications shrink it. The confidence ellipsoid according to the sample mean is given by.

To calculate the first component of the Shahriari et al. (1995) vector to the carbon1 dataset:

```
> p <- 3
> LSL <- c( 0.60, 0.30, 49.00); USL <- c(1.40, 1.70, 51.00); Target <- c(1.00,
  1.00, 50.00)
```

Computing the process region through the proc.reg function for individual observations:

```
> carbon <- matrix(c(carbon1[,1,], carbon1[,2,], carbon1[,3,]),ncol = 3)
> LPL <- proc.reg(carbon, alpha = 0.01)$LPL
> UPL <- proc.reg(carbon, alpha = 0.01)$UPL
```

Computing the process region of the rational subgroups:

```
> x.jk <- apply(carbon1,1:2,mean)
> LPLm <- proc.reg(x.jk, alpha = 0.01)$LPL
> UPLm <- proc.reg(x.jk, alpha = 0.01)$UPL
> Center <- (UPLm + LPLm) /2
```

Then, for proportionality the news specification limits results in:

```
> LSLm <- Target - (Target - LSL) * (Center - LPLm) / (Center - LPL)
> USLm <- Target + (Target - LSL) * (Center - LPLm) / (Center - LPL)
```

Finally the index results in:

```
> CpM <- (prod(USLm - LSLm) / prod(UPL - LPL)) ^ (1 / p)
  1.6547
```

To perform the graphical representation we use the larg.ellip function
The rgl package is required to make the three-dimensional plot.

```
> library(rgl)
> larg.ellip(LSLm, USLm, n = 15, add = FALSE, box = FALSE ,xlim = c
  (0.80,1.150),ylim = c(0.65,1.5), zlim = c(49.5,50.5), xlab = "", ylab = "",
  zlab = "", col = "#D55E00", alpha = 0.2) that builds the largest ellipsoid
  centered at the target.
> plot3d(ellipse3d(cov(x.jk), center = colMeans(x.jk), level = 0.99), type =
  "wire", col = 3, alpha = 0.2, add = TRUE)
```

Afterwards, plot the points

```
> plot3d(x.jk, size = 4, cex = 2, box = FALSE, add = TRUE),
```

and make the cuboids or prisms of the specifications and the modified process region using the prism function.

```
> prism(LSLm, USLm, add = TRUE,col = "#D55E00")
> prism(LPLm, UPLm, add = TRUE,col = 3)
```

The graph obtained allows to visualize it in three dimensions by moving through the axis. In this figure the external prism results in the tolerance region, and the modified tolerance region is represented in the gray ellipsoid. On the other hand, the process region is shown in wire type with its respective modified process region (the external prism that bounds the confidence ellipsoid).

Notice that the first component of the Shahriari et al. (1995) vector is the ratio of both prisms whereas in Taam et al. (1993) the ratio of the ellipsoids always using the volume relative to the tolerances as numerator.

This graph also allows the process monitoring, being the control ellipsoid the χ^2 control chart. In this case no points fall outside the confidence boundaries, then the process seems to be in control (Fig. 3.7).

When the process mean and covariance matrix are known, a it was explained in Chap. 2, then the limits should be replaced. Then, $\frac{p(m-1)(n-1)}{mn-m-p+1}F_{\alpha,p,mn-m-p+1}$ must be substituted by ${\chi_{\alpha,p}}^2$ limits to find the exact limits.

3.6 Multivariate Capability Indices Based on Principal Component Analysis

Many indices based on principal component analysis (PCA) have been proposed in the last years. Some of the most accepted are the indices suggested by Wang and Chen (1998), Xekalaki and Perakis (2002), and Wang (2005). As this approach begins with a PCA, the uncorrelated variables are obtained and the dimensionality reduction is allowed.

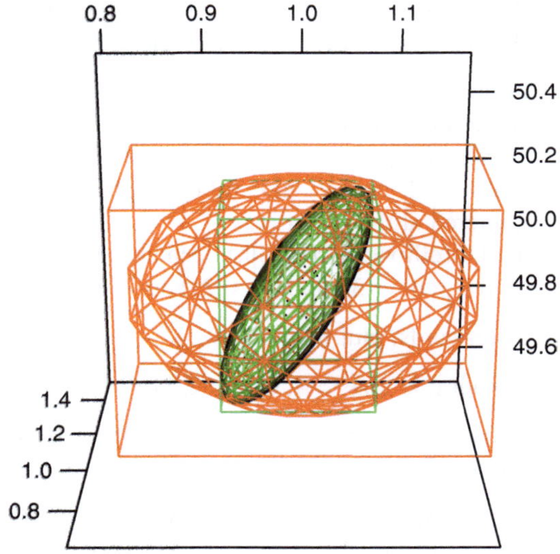

Fig. 3.7 Three-dimensional representation of the carbon1 dataset

These indices are based on the spectral decomposition of the covariance matrix

$$\Sigma = UDU' \tag{3.17}$$

where U is the eigenvectors matrix and D the diagonal matrix of the eigenvalues.

$$D = diag(\lambda_1, \lambda_2, \ldots, \lambda_p) \tag{3.18}$$

The ith principal component results in $PC_i = u_i'x$.

And the engineering specifications (Upper, Lower Specification and Target) are transformed as

$$LSL_{PC_i} = u_i'LSL; \quad USL_{PC_i} = u_i'USL; \quad T_{PC_i} = u_i'T, \tag{3.19}$$

where $i = 1, 2, \ldots, p$ is the ith principal component.

Normally the first components are responsible for most of the variability, therefore the dimensionality can be reduced without significant lost of information. The problem consists on how many components should be retained. In the next section five methods are introduced to deal with this issue.

The proposal by Wang and Chen (1998) is the multivariate extension of the univariate C_p, C_{pk}, C_{pm}, and C_{pmk} indices.

$$MC_p = \left(\prod_{i=1}^{\upsilon} C_{p;PC_i} \right)^{1/\upsilon}, \tag{3.20}$$

where

$$C_{p;PC_i} = \frac{USL_{PC_i} - LSL_{PC_i}}{6\sigma_{PC_i}}, \tag{3.21}$$

where υ is the number of principal component and $\sigma_{PC_i} = \sqrt{\lambda_i}$.

Likewise MC_{pk}, MC_{pm}, and MC_{pmk} are obtained by replacing $C_{p;PC_i}$ by $C_{pk;PC_i}$, $C_{pm;PC_i}$, and $C_{pmk;PC_i}$ respectively, where

$$C_{pk;PC_i} = \min\left\{\frac{USL_{PC_i} - \mu}{3\sigma_{PC_i}}, \frac{\mu - LSL_{PC_i}}{3\sigma_{PC_i}}\right\} \tag{3.22}$$

$$C_{pm;PC_i} = \frac{USL_{PC_i} - LSL_{PC_i}}{6\sqrt{\sigma_{PC_i}^2 + (\mu - T)^2}} \tag{3.23}$$

and,

$$C_{pkm;PC_i} = \frac{(USL_{PC_i} - LSL_{PC_i})/2 - |\mu - [(USL_{PC_i} + LSL_{PC_i})/2]|}{3\sqrt{\sigma_{PC_i}^2 + (\mu - T)^2}} \tag{3.24}$$

Example 3.5

To illustrate the computation of the Wang and Chen (1998) index recall the bimetal1 data introduced in Example 2.9 from the Sect. 2.10. The vectors with the engineering specification are:

$$\begin{aligned} LSL &= [19.0 \quad 39.0 \quad 13.0 \quad 20.2 \quad 24.5] \\ USL &= [23.0 \quad 41.0 \quad 17.0 \quad 23.8 \quad 27.5] \\ \text{Target} &= [21.0 \quad 40.0 \quad 15.0 \quad 22.0 \quad 26.0] \end{aligned}$$

From this dataset the eigenvector and eigenvalues results in:

$$U = \begin{matrix} 0.5968 & 0.5476 & 0.5108 & 0.2881 & 0.0044 \\ 0.2731 & 0.0408 & -0.0739 & -0.5246 & 0.8019 \\ 0.6413 & -0.7344 & -0.0752 & 0.2014 & -0.0561 \\ 0.2988 & 0.3948 & -0.8505 & 0.1404 & -0.1083 \\ 0.2620 & 0.0575 & 0.0673 & -0.7626 & -0.5848 \end{matrix}$$

$$D = 0.1700 \quad 0.0659 \quad 0.0396 \quad 0.0148 \quad 0.0023$$

Consequently the new specifications are given by:

$$\begin{aligned} LSL_{PC_i} &= u_i{}'LSL = [42.78 \quad 11.83 \quad -9.68 \quad -28.21 \quad 14.11] \\ USL_{PC_i} &= u_i{}'USL = [50.14 \quad 12.76 \quad -10.95 \quad -29.09 \quad 13.37] \\ T_{PC_i} &= u_i{}'T = [47 \quad 12 \quad -10 \quad -29 \quad 14] \end{aligned}$$

From Example 2.9, it was determined that the first two principal components were responsible for the 80.61% of the variability. Therefore the dimension of the problem could be reduced to a bivariate alternative.

For the first principal component the Cp index is:

$$C_{p;PC_1} = \frac{USL_{PC_1} - LSL_{PC_1}}{6\sigma_{PC_1}} = \frac{50.14 - 42.78}{6 \times \sqrt{0.1700}} = 2.98$$

And likewise for the other, being:

$$C_{p;PC_2} = 0.60$$

Finally the MC_p is

$$MC_p = \left(\prod_{i=1}^{p} C_{p;PC_i} \right)^{1/p} = (2.98 \times 0.60)^{1/2} = 1.34$$

Using (3.22)–(3.24) the other indices results in:

$$MC_{pk} = 1.13, \quad MC_{pm} = 1.24 \quad \text{and} \quad MC_{pmk} = 1.04$$

The computation in R is based on the mpci function and is as follows:

```
> mpci(index = "wang", bimetal1, LSL, USL, Target, method = 1, perc = 0.80)
```

"Wang and Chen (1998) Multivariate Process Capability Indices (PCI) based on PCA"	$MCp	$MCpm
$'number of principal components'	[1] 1.34	[1] 1.24
[1] 2	$MCpk	$MCpmk
	[1] 1.13	[1] 1.04

As in the index by Wang and Chen (1998) the principal components are taken having the same importance even when the first ones take more weight than the others; Xekalaki and Perakis (2002) proposed to correct that weighting according to the variance explained by the principal components.

$$MXPC_p = \frac{\sum_{i=1}^{\upsilon} \lambda_i C_{p;PC_i}}{\sum_{i=1}^{\upsilon} \lambda_i} \tag{3.25}$$

$MXPC_{pk}$, $MXPC_{pm}$, and $MXPC_{pmk}$ are similarly achieved.

The calculation of the Xekalaki and Perakis (2002) index is presented in the following example.

Example 3.6
Using the same engineering specifications and number of the principal components in the bimetal1 dataset

$$MXPC_p = \frac{\sum_{i=1}^{2} \lambda_i C_{p;PC_i}}{\sum_{i=1}^{2} \lambda_i} = \frac{0.17 \times 2.98 + 0.07 \times 0.6}{0.17 + 0.07} = 2.31$$

$$MXPC_{pm} = 2.17; \quad MXPC_{pk} = 2.18 \quad \text{and} \quad MXPC_{pmk} = 2.05$$

The use of the function mpci in this context is as follows:

```
> mpci(index = "xeke", bimetal1, LSL, USL, Target, method = 1, perc = 0.80)
```

"Xekalaki and Perakis (2002) Multivariate Process Capability Indices (PCI) based on PCA"	$MCp	$MCpm
$'number of principal components'	[1] 2.31	[1] 2.17
[1] 2	$MCpk	$MCpmk
	[1] 2.18	[1] 2.05

On the other hand, Wang (2005) suggests another way to weight the principal components using the weighted geometric mean. The proposed indices result in:

$$MWC_p = \left(\prod_{i=1}^{v} C_{p;PC_i}^{\lambda_i} \right)^{1/\sum_{i=1}^{v} \lambda_i} \tag{3.26}$$

and so on for MWC_{pk}, MWC_{pm}, and MWC_{pmk}.

Example 3.7
This example computes the indices according to the Wang's (2005) method.

$$MWC_p = \left(\prod_{i=1}^{2} C_{p;PC_i}^{\lambda_i} \right)^{1/\sum_{i=1}^{2} \lambda_i} = \left(2.98^{0.17} \times 0.6^{0.07}\right)^{1/(0.17+0.07)} = 1.90$$

$$MWC_{pk} = 1.70; \quad MWC_{pm} = 1.77; \quad MWC_{pmk} = 1.58$$

To perform these indices in R just use the argument index = "wangw" in mpci function.

```
> mpci(index = "wangw", bimetal1, LSL, USL, Target, method = 1, perc = 0.80)
```

"Wang (2005) Multivariate Process Capability Indices(PCI) based on PCA"	$MCp	$MCpm
$ 'number of principal components'	[1] 1.91	[1] 1.77
2	$MCpk	$MCpmk
	[1] 1.70	[1] 1.58

3.7 Methodology to Select the Number of Principal Components

In previous sections, it was tackled how the principal components analysis allow the dimensionality reduction of the data in which $1 \leq \lambda \leq p$ principal components can be obtained. There are many methods in order to decide how many principal components should be retained or used, with the aim to avoid the loss of significant information.

Rencher (2002) proposed the next four methods and we add a fifth.
Method 1 or Percentage: This technique guarantees at least the percent specified of Cumulative Proportion of explained variance. This is normally fixed on 80%.

Example 3.8
In Example 2.5 a dataset called bimetal1 collected from a certain type of strip composed of brass and steel with five quality characteristics and 28 samples was introduced.

Using summary (princomp(bimetal1)) R shows a summary that includes the standard deviation, the proportion of variance explained, and the cumulative proportion of the eigenvalues (Table 3.1).

If the threshold of the 80% is used then the first two components should be retained.

Method 2 or Average: The second method is based on retaining the principal components whose eigenvalues are greater than the average of the eigenvalues.

$$\sum_{i=1}^{p} \lambda_i/p$$

The eigenvalues are easily computed:

```
eig <- eigen(cov(bimetal1))$values; print(eig)
[1] 0.169984728 0.065883347 0.039640343 0.014847291 0.002264529
```

If mean(eig)=0.05852405, therefore only the first two components comply with this condition.

Method 3 or Scree: The scree graph is a visual procedure that plots the eigenvalue size throughout the eigenvalue number. It allows to determine which components are significant apart from the straight line formed by the last eingenvalues.

In the example the scree graph shows that the first component is separated from the straight line, therefore the first principal component should be retained (Fig. 3.8).

Table 3.1 Importance of components

	Comp.1	Comp.2	Comp.3	Comp.4	Comp.5
Standard deviation	0.40486	0.25205	0.19551	0.11965	0.04673
Proportion of variance	0.58091	0.22515	0.13547	0.05074	0.00774
Cumulative proportion	0.58091	0.80606	0.94152	0.99226	1.00000

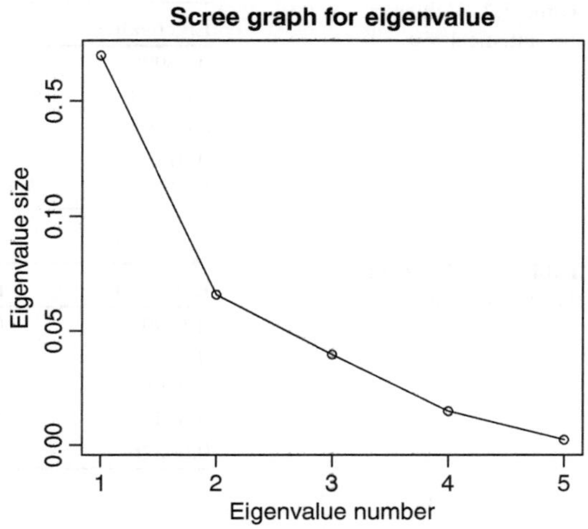

Fig. 3.8 Scree graph for the eigenvalues

Method 4 or Bartlett (1954) Test: This method is a statistical test designed to ignore the principal components not significantly different from the rest and assumes multivariate normality. Usually this method produces a number of principal components larger than the former methods. For more details see the following example: Rencher (2002)

$$H_0 : \lambda_1 = \lambda_2 = \ldots = \lambda_p$$
$$H_1 : \lambda_i \neq \lambda_j \quad \text{for some} \quad i \neq j$$

$$\overline{\lambda} = \sum\nolimits_{i=p-k+1}^{p} \lambda_i / k \tag{3.27}$$

where k is the sequence p, p − 1, p − 2, . . ., 1

$$\chi^2 = \left(n - \frac{2p+11}{6}\right)\left(k \ln \overline{\lambda} - \sum_{i=p-k+1}^{p} \ln \lambda_i\right) \tag{3.28}$$

$$\chi^2 \geq \chi^2_{\alpha,1/2(k-1)(k+2)} \tag{3.29}$$

The results of the practical and theoretical χ^2s are (Table 3.2):

This implies that the first four are significantly different from each other. Therefore according to the Bartlett's Test in this case the first four principal components should be retained.

Method 5 or Anderson (1963) Test: Another method widely used is the Anderson Test that differentiates also the principal components significantly different from the others.

Table 3.2 Values of the statistical test

Eigenvalue	k	χ^2	$\chi^2_{\alpha,1/2(k-1)(k+2)}$
0.16998	5	93.80	33.20
0.06588	4	56.57	25.26
0.03964	3	39.82	18.21
0.01485	2	19.06	11.83
0.00226	1	0	0

Table 3.3 Values of the statistical test.

Eigenvalue	k	χ^2	$\chi^2_{\alpha,1/2(k-1)(k+2)}$
0.16998	0	103.37	33.20
0.06588	1	62.34	25.26
0.03964	2	43.88	18.21
0.01485	3	21.01	11.83
0.00226	4	0	0

$$H_0 : \lambda_1 = \lambda_2 = \ldots = \lambda_p$$
$$H_1 : \lambda_i \neq \lambda_j \quad \text{for some} \quad i \neq j,$$

where k = 1, 2, . . ., p

$$\chi^2 = -\upsilon \sum_{i=k+1}^{p} \ln(\lambda_i) + \upsilon(p-k)\ln\left(\frac{\sum_{i=k+1}^{p} \lambda_i}{p-k}\right) \tag{3.30}$$

$$\chi^2 \geq \chi^2_{\alpha,1/2(p-k-1)(p-k+2)} \tag{3.31}$$

The results are shown in Table 3.3:

This method found the first four eingenvalues significantly different, as a result the first four must be retained.

3.8 Exercises

3.1. The indust1 dataset represents the data obtained from an industrial process in which two correlated quality characteristics are controlled. The engineering tolerances are: $LSL_1 = 2.8$, $LSL_2 = 5.5$, $USL_1 = 5.5$, and $USL_2 = 8.7$. Use alpha = 0.0027.

(a) Compute the Shahriari et al. (1995) vector. Interpret the result of each component of the vector.

(b) Determine the capability index MCpm according to Taam et al. (1993). Compare the result with the first component of the Shahriari et al. (1995) vector.

(c) Compute the Pan and Lee (2010) index and contrast it with the two previous indices.
(d) Compare the Taam et al. (1993) index using alpha = 0.001 with the achieved in clause (b).

3.2. The dataset called water1 consists on five variables (pH, phosphates (mg/L), nitrates (mg/L), dissolved oxygen, and total solids (mg/L)) measured in a water quality test. For all clauses consider alpha = 0.001. The following vectors represent the specifications: LSL = [3.00, 0.01, 0.01, 88.00, 145.00] and USL = [11.00, 0.50, 1.30, 110.00, 200.00].

(a) Compute the correlation matrix.
(b) Compare the Taam et al. (1993) with the Pan and Lee (2010) index.
(c) Determine if the modified process region is contained by the tolerance region.
(d) Assess the closeness of the process mean with the tolerance target value.
(e) Compute the capability indices according to Wang and Chen (1998) using the method = 1 to select the number of principal components.
(f) Compare the values achieved using the two first principal components in the Wang (2005) and Xekalaki and Perakis (2002) indices.
(g) According to the Scree graph. How many principal components should be retained?

3.3. The dataset mech1 represents the data obtained from seven quality characteristics collected from a mechanical process. Use alpha = 0.0027 and the following vectors are the engineering specifications:

LSL = [5.00, 33.00, 3.50, 3.00, 1.00, 37.00, 118.00].
USL = [15.00, 37.00, 6.50, 17.00, 41.00, 43.00, 122.00].

(a) Compute the Shahriari et al. (1995) vector.
(b) Assess the closeness between the process mean and the engineering target value.
(c) Determine if the modified process region is contained by the tolerance region.
(d) Compute the Wang (2005) indices using the Scree graph.
(e) Determine the Xekalaki and Perakis (2002) indices using method 4 (Bartlett Test). How many principal components were retained?
(f) Is this last result significantly different if two principal components are retained?
(g) Compute both Taam et al. (1993) and Pan and Lee (2010) indices. Discuss the results.

3.4. Three variables were collected and stored in glass1 dataset to develop a capability study. The dataset is presented in rational subgroups because it was gathered initially to perform a multivariate process monitoring program. To transform it to a 2D array or matrix use.

```
glass <- matrix(c(glass1[,1,], glass1[,2,], glass1[,3,]),ncol = 3)
```

The specifications for each variable are defined by the interval: [9.00, 11.00], [0.50, 3.50], and [3.5, 6.50] respectively.

(a) Compute the three indices based on tolerance and process region ratios (Shahriari et al. 1995; Taam et al. 1993; Pan and Lee 2010). Compare the results.
(b) Calculate the indices based on PCA. Discuss the results achieved.

3.5. In the previous chapter it was studied a manufacturing process of certain type of carbon tubing, composed by three quality characteristics and in Sect. 3.4 it was studied as a rational subgroup case. The rational subgroup can be eliminated using

```
carbon <- matrix(c(carbon1[,1,], carbon1[,2,], carbon1[,3,]),ncol = 3)
```

obtaining a 2D array. In this process the specifications are given by:
LSL = [0.60, 0.30, 49.00]
USL = [1.40, 1.70, 51.00]
Target = [1.00, 1.00, 50.00]

(a) Compare the Pan and Lee (2010) NMCpm index with the first component of the Shahriari et al. (1995) vector using alpha = 0.0001.
(b) Contrast both indices but setting the midpoint between specifications as Target.

3.6. Consider the first two quality characteristics of the water1 dataset and the following specifications:
$LSL_1 = 3.00$, $LSL_2 = 0.01$, $USL_1 = 10.5$ and $USL_2 = 0.45$. The Target of the process is given by: $T_1 = 7.00$, $LSL_2 = 0.23$ Use alpha = 0.001.

(a) Calculate the Shahriari et al. (1995) vector. Interpret the result of each component of the vector and obtain the graphical representation. Explain the result graphically.
(b) Compute the capability index MCpm according to Taam et al. (1993) and the NMCpm by Pan and Lee (2010). Explain graphically how both indices are computed.

3.7. Which of the following indices is computed as the ratio of the largest ellipsoid centered at the target to the process region?

- Wang (2005) indices
- Taam et al. (1993) MCpm
- Shahriari et al. (1995) first component.

3.8. Reconsider the dataset called mech1 to the exercise 3.3. Exclude the variables x_1 and x_6 using:

```
mech <- mech1[,c(−1,-6)].
```

(a) Determine if the modified process region is contained by the tolerance region using alpha = 0.0002.
(b) Estimate the Pan and Lee (2010) index.
(c) Calculate the capability indices according to Wang (2005) using the method = 2 to select the number of principal components.
(d) Compare the values achieved using the two first principal components in the Wang and Chen (1998) and Xekalaki and Perakis (2002) indices.

Chapter 4
Tools of Support to MSQC

4.1 Tools of Support to MSQC

As a general rule, normality and independence of the data is required in Statistical Process Control and the multivariate extensions are not the exception. In a multivariate control chart with the use of rational subgroups according to the central limits theorem certain grade of normality is achieved. But in alternatives called charts for individuals, this rule is not satisfied. The same occurs in capability indices that rarely are computed using subgroups.

Many authors have proposed nonparametric alternatives to deal with the departures of normality and techniques based on PCA as the studied in Sects. 2.10 and 3.6 which are robust to the lack of normality.

However, nowadays it results quite unproblematic to test multivariate normality and randomness. In this chapter we introduce a wide range of tools to fulfill these requirements.

4.1.1 Graphical Methods

The first section of this chapter will examine two graphical techniques: histogram and Q-Q plot that facilitate the assumption of normality.

Histogram is a graphical technique that allows a visual summary of the data. It provides information about the center, the spread, the skewness, and the existence of outliers. (NIST / SEMATECH e-Handbook of Statistical Methods).

A visual inspection of a histogram permits to establish an initial hypothesis of the distribution; in this case a bell-shaped is desired.

Although histograms are basically used in univariate scenarios, univariate normality per se does not imply multivariate normality; if a departure from normality is founded in individual variables, this has a negative effect in the multinormality.

E. Santos-Fernández, *Multivariate Statistical Quality Control Using R*,
SpringerBriefs in Statistics 14, DOI 10.1007/978-1-4614-5453-3_4,

Example 4.1
In this example we will illustrate the use of histogram in a multivariate context. For that, return to the bimetal dataset introduced in Sect. 2.6.

To put multiple figures in one graph device the parameter mfrow can be used by specifying mfrow = c(n,m) being n the number of figures by row and m by columns.

```
> par(mfrow = c(3,2))
```

As for each quality characteristic a histogram is desired—a simple loop is used.

```
> for( i in 1 : ncol(bimetal1) ){
> x <- bimetal1[,i]
> mean<-mean(bimetal1[,i])
> sd<-sd(bimetal1[,i])
> hist(x, prob = TRUE, main = paste( "Histogram for ", colnames(bimetal1)[i] ),
  xlab = "")
```

Finally, adding the normal curve

```
> points(curve(dnorm(x, mean = mean, sd = sd), add = TRUE),type = "l")}
```

From this chart we can appreciate that most of the classes are located in the center, no significant skewness is revealed, no long tails are presented, and no considerable outliers are detected. The form of the classes does not differ drastically to the normal shape. Finally, there is no visual evidence to reject the univariate normality hypothesis.

This visual inspection can be complemented with the quantile-quantile plot, or simply Q-Q plot.

The Q-Q plot is a graphical tool for comparing a two dataset or a dataset with a theoretical distribution. The most common use is to plot the quantiles against a

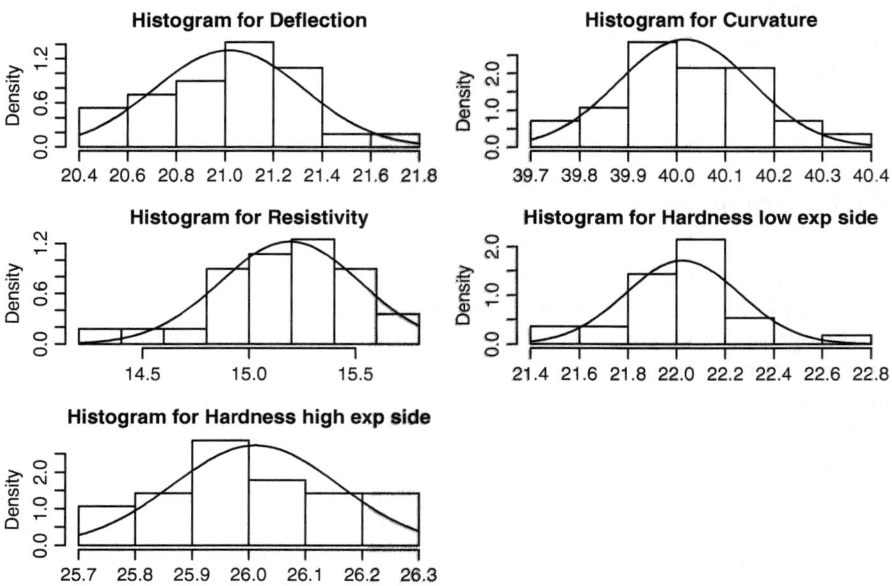

Fig. 4.1 Histogram of the individual variables in the bimetal1 dataset

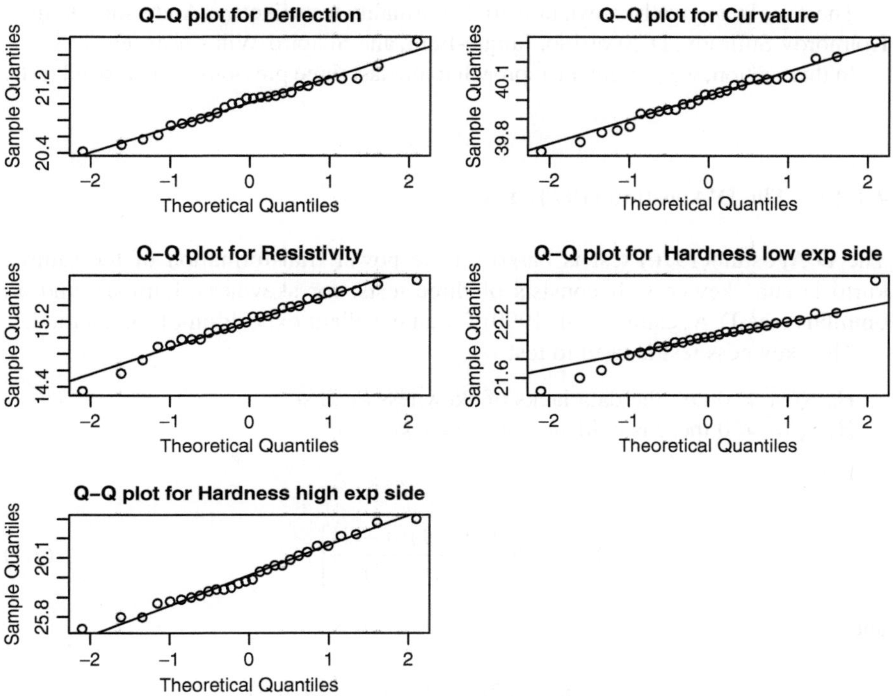

Fig. 4.2 The Q-Q plot of the individual variables in bimetal1 dataset

reference line from a normal distribution. When the points fall approximately over the line there is evidence that both come from an identical distribution.

The performing of a Q-Q plot in R is done through the qqnorm function.

Example 4.2
To construct a Q-Q for each variable from bimetal1 dataset:

```
> par(mfrow = c(3,2))
> for( i in 1 : ncol(bimetal1) ){
> qqnorm(bimetal1[,i], main = paste( "Q-Q plot for ", colnames(bimetal1)[i] ) )
```

And to include the reference line from the normal distribution

```
> qqline(bimetal1[,i])
> }
```

From these graphs it appears that each variable is normally distributed since no departure from diagonal line is presented (Fig. 4.2).

4.1.2 Marginal Normality Test

Although in a p-variate data the marginal normality does not imply joint normality, deviation from normality frequently affects the marginal distributions.

There are many well-known univariate normality tests like: χ^2, Anderson-Darling, Kolmorov-Smirnov, D'Agostino, Jarque-Bera, and Shapiro-Wilks tests, etc.

In this section, we present an approach to the last three previously mentioned tests.

4.1.2.1 The D'Agostino (1970) Test

The D'Agostino(1970) test is based on the power transformation of the sample kurtosis and skewness. It consists of three tests: for skewness, kurtosis, and an omnibus (see D'Agostino et al. 1990) for an excellent exposition of the method.

The skewness test is used to test

H_o: $\sqrt{b_1} = 0$ i.e.: the data lacks of skewness against
H_1: $\sqrt{b_1} \neq 0$ there is evidence of skewness.

Let

$$Y = \sqrt{b_1}\left[\frac{(n+1)(n+3)}{6(n-2)}\right]^{1/2} \tag{4.1}$$

and

$$B = \frac{3(n^2+27n-70)(n+1)(n+3)}{(n-2)(n+5)(n+7)(n+9)} \tag{4.2}$$

Where n is the sample size. Using the Johnson's unbounded (SU) the $X(\sqrt{b_1})$ has a normal distribution, being:

$$X\left(\sqrt{b_1}\right) = \delta \log\left(\frac{y}{\alpha} + \sqrt{\left(\frac{y}{\alpha}\right)^2 + 1}\right) \tag{4.3}$$

where δ and α are determined as:

$$\delta = \frac{1}{\sqrt{\log(W)}} \tag{4.4}$$

and

$$\alpha = \sqrt{\frac{2}{W^2 - 1}} \tag{4.5}$$

with

$$W^2 = \sqrt{2(B-1)} - 1 \tag{4.6}$$

The kurtosis test is based on the following hypothesis

H_o: $b_2 = 3$ and H_1: $b_2 \neq 3$

The mean and variance are computed as follows:

$$E(b_2) = \frac{3(n-1)}{n+1} \tag{4.7}$$

and

$$\operatorname{var}(b_2) = \frac{24m(m-2)(m-3)}{(m+1)^2(m+3)(m+5)} \tag{4.8}$$

Then standardizing b_2

$$x = \frac{b_2 - E(b_2)}{\sqrt{\operatorname{var}(b_2)}} \tag{4.9}$$

and calculating the statistics

$$Z(b_2) = \frac{\left(1 - \frac{2}{9A}\right) - \left[\frac{1-2/A}{1+x\sqrt{2/(A-4)}}\right]^{1/3}}{\sqrt{2/9A}} \tag{4.10}$$

where

$$A = 6 + \frac{8}{\sqrt{\beta_1(b_2)}}\left[\frac{2}{\sqrt{\beta_1(b_2)}} + \sqrt{\left(1 + \frac{4}{\beta_1(b_2)}\right)}\right] \tag{4.11}$$

and

$$\beta_1(b_2) = \frac{6(m^2 - 5m + 2)}{(m+7)(m+9)}\sqrt{\frac{6(m+3)(m+5)}{m(m-2)(m-3)}} \tag{4.12}$$

The $Z(b_2)$ statistics has approximately a normal distribution

4.1.2.2 Omnibus Test

In order to integrate both tests, D'Agostino and Pearson (1973) proposed an omnibus test with the following statistics

$$K^2 = Z^2\left(\sqrt{b_1}\right) + Z^2(b_2) \tag{4.13}$$

Table 4.1 Results of the D'Agostino Test for each variable of the bimetal dataset

D'Agostino Test for the deflection Skewness Skewness coefficient: 0.08 Statistics: 0.21 p-value: 0.83 Kurtosis The kurtosis coefficient: 3.04 Statistics: 0.59 p-value: 0.56 Omnibus Test Chi-squared: 0.39 Degree of freedom: 2 p-value: 0.82	D'Agostino Test for the curvature Skewness Skewness coefficient: -0.07 Statistics: -0.18 p-value: 0.85 Kurtosis The kurtosis coefficient: 2.75 Statistics: 0.17 p-value: 0.86 Omnibus Test Chi-squared: 0.06 Degree of freedom: 2 p-value: 0.97	D'Agostino Test for the resistivity Skewness coefficient: -0.61 Statistics: -1.5 p-value: 0.13 Kurtosis The kurtosis coefficient: 3.14 Statistics: 0.71 p-value: 0.47 Omnibus Test Chi-squared: 2.76 Degree of freedom: 2 p-value: 0.25
D'Agostino Test for the low expansion side Skewness Skewness coefficient: -0.04 Statistics: -0.11 p-value: 0.92 Kurtosis The kurtosis coefficient: 4.16 Statistics: 1.67 p-value: 0.09 Omnibus Test Chi-squared: 2.81 Degree of freedom: 2 p-value: 0.25	D'Agostino Test for the high expansion side Skewness Skewness coefficient: 0.23 Statistics: 0.58 p-value: 0.56 Kurtosis The kurtosis coefficient: 2.29 Statistics: -0.71 p-value: 0.48 Omnibus Test Chi-squared: 0.85 Degree of freedom: 2 p-value: 0.66	

where K^2 follows a χ^2 distribution with two degrees of freedom.

$$K^2 \sim \chi^2_{\alpha,2} \tag{4.14}$$

Example 4.3
To illustrate the use of the D'Agostino test in **R**, use the function DAGOSTINO included in the MSQC package, as follows:

```
> for (i in 1 : 5){
> DAGOSTINO(bimetal1[,i])}
```

The Table shows the results achieved (Table 4.1)

As a result of the Skewness Test, no significant lack of symmetry is presented. Certain grade of skewness is obtained in the resistivity though, corroborating the result obtained for the histogram in Fig. 4.1.

Conversely, the Kurtosis Test detects a positive grade of peakness in a low expansion side variable since the kurtosis coefficient was 4.16 although not significant at alpha = 0.05 (see p-value: 0.09)

On the other hand the omnibus test does not found departures from normality.

According to this test, there is no evidence for rejecting the normality assumption.

4.1.2.3 The Jarque and Bera (1980) Test

The Jarque and Bera (1980) Test is an elegant and powerful goodness of fit test, likewise based on kurtosis and skewness. It is defined as:

$$JB = \frac{m}{6}\left[S^2 + \frac{1}{4}(K-3)^2\right] \tag{4.15}$$

where m is the sample size and S and K the skewness and kurtosis respectively.

The JB statistics follows a χ^2 distribution with two degrees of freedom.

For more details see Jarque and Bera (1980), Jarque and Bera (1987), or Jarque (2010).

Jarque (2010) offers the significance points table although statistical software usually computes the p-values as:

p-value=1 - pchisq(STATISTIC, df = 2) or p-value = $1 - \chi^2{}_{JB,2}$

At least three R packages include this test. They are: tseries, moments, and lawstat.

In this context we use the first one:

```
> library("tseries")
```

Example 4.4
Using the jarque.bera.test function for each quality characteristics from the bimetal1 dataset:

Jarque-Bera Test data: bimetal1[, 1] X-squared = 0.22, df = 2, p-value = 0.90	Jarque-Bera Test data: bimetal1[, 3] X-squared = 1.87, df = 2, p-value = 0.39	Jarque-Bera Test data: bimetal1 [, 5] X-squared = 0.57, df = 2, p-value = 0.75
Jarque Bera Test data: bimetal1[, 2] X-squared = 1.74, df = 2, p-value = 0.42	Jarque Bera Test data: bimetal1[, 4] X-squared = 0.96, df = 2, p-value = 0.62	

Notice that according to the p-values the normality assumption cannot be rejected at alpha level = 0.05 or 0.10.

4.1.2.4 The Shapiro and Wilk (1965) Test

The Shapiro and Wilk (1965) Test has become one of the most popular tests due to its high performance.

The null hypothesis H_0 is the sample that proceeds from a normal distribution and possesses the statistics

$$W = \frac{\left[\sum_1^m a_i x_{(i)}\right]^2}{\sum_1^m x_i - \bar{x}} \tag{4.16}$$

where

$$a' = (a_1, a_2, ..., a_m) = w'V^{-1}\left[\left(w'V^{-1}\right)\left('V^{-1}w\right)\right]^{-1/2} \tag{4.17}$$

and w the normal scores and V its covariance matrix.

They proposed the approximation of a' as:

$$\hat{a}_i' = \begin{cases} 2w_k & 1<k<m \\ \left(\frac{\hat{a}_1^2}{1-2\hat{a}_1^2}\sum_{k=2}^{m-1}\hat{a}_1^2\right)^{1/2} & i=1,m \end{cases} \tag{4.18}$$

where

$$\hat{a}_1' = \hat{a}_m' = \begin{cases} g(m-1) & m \le 20 \\ g(m) & m>20 \end{cases} \tag{4.19}$$

being

$$g(m) = \frac{\Gamma\left[\frac{1}{2}(m+1)\right]}{\sqrt{2}\Gamma\left(\frac{1}{2}m+1\right)} \tag{4.20}$$

Using the approximation:

$$g(m) = \left(\frac{6m+7}{6m+13}\right)\left[\frac{\exp(1)}{m+2}\left(\frac{m+1}{n+2}\right)^{n-2}\right]^{1/2} \tag{4.21}$$

Royston (1982) proposed the transformation of W for $7 \le m \le 2000$ to normality as follows:

$$x = (1-W)^{\lambda} \tag{4.22}$$

and

$$z = \frac{(x - \mu_x)}{\sigma_x} \tag{4.23}$$

R includes the built-in function shapiro.test() to compute this test.

Example 4.5
The example below illustrates its use over the bimetal1 dataset. Using this function individually for every quality characteristic

Shapiro-Wilk normality test data: deflection W = 0.98, p-value = 0.86	Shapiro-Wilk normality test data: curvature W = 0.98, p-value = 0.89
Shapiro-Wilk normality test data: resistivity W = 0.97, p-value = 0.46	Shapiro-Wilk normality test data: low expansion side W = 0.97, p-value = 0.46
Shapiro-Wilk normality test data: high expansion side W = 0.98, p-value = 0.78	

On the other hand, Thode (2010) offers an excellent presentation of the most powerful test and suggests a test based on moments like Shapiro-Wilks, Anderson-Darling, and Jarque Bera. For more details see Thode (2002).

4.1.3 Assessing Multivariate Normality

Though the literature reflects that the proposals to test multivariate normality exceed the 50 methods (see e.g.: (Mecklin and Mundfrom 2004)) these tools are rarely applied in MSPC publications. This is due to the fact that as a general rule these methods lack of simplicity and the software availability is limited.

Three of the most powerful tests are introduced in this section.

4.1.3.1 Mardia (1970) Skewness and Kurtosis Test

The Mardia (1970) test is a generalization of the univariate skewness and kurtosis test and becomes one of the most popular ones on assessment of multivariate normality. The multivariate skewness and kurtosis are given by:

$$b_{1,p} = \frac{1}{n^2} \sum_{i=1}^{n} \sum_{j=1}^{n} g_{jk}^3 \tag{4.24}$$

$$b_{2,p} = \frac{1}{n} \sum_{i=1}^{n} g_{jj}^2 \tag{4.25}$$

1 where:

$$g_{jk}^3 = \left[(x_j - \bar{x})'\Sigma^{-1}(x_k - \bar{x})\right]^3 \tag{4.26}$$

and

$$g_{jj}^2 = \left[(x_j - \bar{x})'\Sigma^{-1}(x_j - \bar{x})\right]^2 \tag{4.27}$$

Mardia (1970, 1974) provides the percentiles for $b_{1,p}$ and $b_{2,p}$ for many values of p (quality characteristics) and many numbers of samples (m).

Mardia also proposed for $b_{1,p}$ an approximation to the χ^2 distribution as follows:

$$b_{1,p}\frac{(p+1)(m+1)(m+3)}{6[(m+1)(p+1)-6]} \sim \chi^2_{\alpha,[p(p+1)(p+2)]/6} \tag{4.28}$$

while for $b_{2,p}$ a normal approximation, being:

$$b_{2,p} \sim N(p(p+2), 8p(p+2)/m) \tag{4.29}$$

The Mardia test is available from QRMlib and dprep R packages.

Example 4.6
Then, to illustrate the Mardia Test return to the bimetal1 dataset.

Using the QRMlib package:

```
> MardiaTest(bimetal1)
```

The R returns

```
$skewness
[1] 6.982112
$p.value
[1] 0.585327
$kurtosis
[1] 33.77373
$p.value
[1] 0.3490892
```

Regarding the p-value for skewness and kurtosis, there is no evidence of departures from normality.

4.1.3.2 Henze and Zirkler (1990) Test

Henze and Zirkler (1990) proposed a multivariate normality test based on the empirical characteristic function. A wide number of simulation studies point out the high performance of this test. See e.g.: (Thode 2002)

The statistics is given by:

$$T = \frac{1}{n^2}\sum_{k}^{p}\sum_{j=1}^{m} e^{-\frac{b^2}{2}\left|y_j - y_k\right|^2} + \left(1+2b^2\right)^{-m/2} - 2\left(1+b^2\right)^{-m/2}\frac{1}{n}\sum_{j=1}^{m} e^{-\frac{b^2}{2\left(1+b^2\right)}y_j^2} \tag{4.30}$$

where

$$\left|y_j - y_k\right|^2 = \left(x_j - x_k\right)' S^{-1}\left(x_j - x_k\right) \tag{4.31}$$

$$y_j{}^2 = \left(x_j - \overline{x}\right)' S^{-1}\left(x_j - \overline{x}\right) \tag{4.32}$$

and

$$b = \frac{1}{\sqrt{2}}\left[\frac{n(2m+1)}{4}\right]^{1/(m+4)} \tag{4.33}$$

T has a lognormal distribution with mean

$$\overline{T} = 1 - \left(1+2b^2\right)^{-m/2}\left(1 + \frac{mb^2}{1+2b^2} + \frac{m(m+2)b^4}{2(1+2b^2)^2}\right) \tag{4.34}$$

and variance

$$\begin{aligned}\mathrm{var}(T) = {} & 2\left(1+4b^2\right)^{-m/2} + 2\left(1+2b^2\right)^{-m}\left[1 + \frac{2mb^4}{\left(1+2b^2\right)^2} + \frac{3m(m+2)b^8}{4\left(1+2b^2\right)^4}\right] \\ & - 4w^{-m/2}\left[1 + \frac{3mb^4}{2w} + \frac{m(m+2)b^8}{2w^2}\right]\end{aligned} \tag{4.35}$$

where

$$w = \left(1+b^2\right)\left(1+3b^2\right) \tag{4.36}$$

$T \sim L_{\mu,\sigma}$ where

$$\mu = \log\left[\left(\frac{\overline{T}^4}{\mathrm{var}(T) + \overline{T}^2}\right)^{1/2}\right] \tag{4.37}$$

$$\sigma = \left[\log\left(\frac{\mathrm{var}(T) + \overline{T}^2}{\overline{T}^2}\right)\right]^{1/2} \tag{4.38}$$

The HZ.test function is available on the MSQC package

Example 4.7

The following example shows the application of the test using also the bimetal1 data:

```
> HZ.test(bimetal1)
p-value    HZ statistic
[1] 0.61   0.77
```

According to the results obtained, p-value = 0.77, which is a high value; there is no evidence to reject the assumption of multivariate normality.

4.1.3.3 Royston (1992)Test

Another powerful test was proposed by Royston (1983) which is a multivariate extension of the Shapiro and Wilks normality test (see Royston 1982, 1983, 1992, 1995).

The statistic recommended by Royston is

$$H = \frac{e\sum_{j=1}^{p} R_j'}{p} \tag{4.39}$$

where

$$R_j = \left\{\Phi^{-1}\left[\frac{\Phi(-Z_j)}{2}\right]\right\}^2 \tag{4.40}$$

There are two ways to compute Z_j according to the number of observations:
For $4 \le n \le 11$

$$Z_j = \frac{\log\{\gamma - [\log(1 - W_j)]\} - \mu}{\sigma} \tag{4.41}$$

W_j is the statistics of the univariate Shapiro-Wilks test. (See the previous section.)

where

$$\gamma = -2.273 + 0.459n \tag{4.42}$$

$$\mu = 0.544 - 0.39978n + 0.025054n^2 - 0.0006714n^3 \tag{4.43}$$

$$\sigma = \exp(1.3822 - 0.77875n + 0.062767n^2 - 0.0020322n^3) \tag{4.44}$$

and for $12 <= x <= 2000$

$$Z_j = \frac{\log(1 - W_j) + \gamma - \mu}{\sigma} \tag{4.45}$$

where $\gamma = 0$

$$\mu = -1.5861 - 0.31802\log(n) + 0.083751[\log(n)]^2 + 0.0038915[\log(n)]^3 \tag{4.46}$$

$$\sigma = \exp\left(-0.4803 - 0.082676 + 0.062767[\log(n)]^2 - 0.0030302[\log(n)]^3\right) \tag{4.47}$$

Otherwise, e in the H statistics is given by:

$$e = \frac{m}{1 + (m - 1)\overline{c}} \tag{4.48}$$

Where

$$\overline{c} = \frac{\sum_{j=1}^{p}\sum_{k=1}^{m} c_{ij}^5 - m}{m^2 - m} \tag{4.49}$$

and

$$c_{ij}^5 = r_{ij}^5\left[1 - \frac{0.715\left(1 - r_{ij}\right)^{0.715}}{v}\right] \tag{4.50}$$

being r_{ij} the correlation and

$$v = 0.21364 + 0.015124\log^2(n) - 0.0018034\log^3(n) \tag{4.51}$$

Royston's H statistics follow approximately a χ^2 distribution with e degrees of freedom.

This function is also included in the MSQC package and the usage is as follows

Example 4.8

> Royston.test(bimetal1)
Then R prompts:
test.statistic p.value
1.18 0.94

With the p-value obtained, there is no evidence of departure from multivariate normality at significance level of 0.05.

4.1.4 Solutions to Departures from Normality

Practically, it is common to get variables with non-normal distribution and one alternative is to transform the data. The transformation of the data is the application of a mathematical function to the original dataset.

In a multivariate context this solution could be addressed to a marginal or multivariate approach. In this section two marginal solutions and one multivariate are introduced.

There are many simple transformations used in practice: $\sqrt{x}$, log(x), arcsin($\sqrt{x}$), etc (see, e.g., (Juran and Godfrey 1998) Sect. 4.4)

Another is the well-known Box-Cox Transformation (BCT) that is probably the most used one for practitioners and professionals of quality control. Finally, another type of transformation (although not so well known) is the Johnson's system of distributions recognized as the Johnson Transformation (JT)

4.1.4.1 Box-Cox Transformation (BCT)

The family of Box-Cox is a power transformation suggested by Box and Cox (1964). It is given by:

$$y_i = \begin{cases} \dfrac{x_i^{\lambda} - 1}{\lambda} & for \quad \lambda \neq 0 \\ \log(x_i) & for \quad \lambda = 0 \end{cases} \tag{4.52}$$

where x_i is the original dataset, λ (lambda) is the power and y_i the new observations. One alternative, in order to find the optimal value of λ, is using the value that maximizes the logarithm of the likelihood function. For more details see Box and Cox (1964) or Venables and Ripley (2002).

The BCT is widely used to improve the normality in some practical situations and a lot of statistical packages provide this application. An advantage is the easy algorithm to transform the data while a disadvantage is that it does not allow negative data values, though it can be solved by adding a constant to the original dataset.

There are many functions in R that perform the BCT transformation but we will use the powerTransform included in car package.

4.1.4.2 Johnson Transformation (JT)

The Z family of distributions was presented in Johnson (1949) and is composed by three distributions named Unbounded (SU), Lognormal (SL), and Bounded (SB) which allow to transform into a normal distribution through selecting one of the three of them. The transformations are:

$$SU \quad Z = \gamma + \eta \sinh^{-1}\left(\frac{x - \varepsilon}{\lambda}\right) \tag{4.53}$$

where

$\eta, \lambda > 0, \; -\infty < \gamma < \infty, \; -\infty < \varepsilon < \infty$, and $-\infty < x < \infty$

$$SL \quad Z = \gamma + \eta \ln^{-1}(x - \varepsilon) \tag{4.54}$$

where

$\eta > 0, \; -\infty < \gamma < \infty, \; -\infty < \varepsilon < \infty$ and $\varepsilon < x$

$$SB \quad Z = \gamma + \eta \ln\left(\frac{x - \varepsilon}{\lambda + \varepsilon - x}\right) \tag{4.55}$$

where

$\eta, \lambda > 0, \; -\infty < \gamma < \infty, \; -\infty < \varepsilon < \infty$ and $\varepsilon < x < \varepsilon + \lambda$

Chou et al. (1998) proposed a methodology to transform non-normal data using the method of percentiles distribution. The method optimizes the transformation based on the parameter estimates suggested by Slifker and Shapiro (1980), finding the best fit to the standard normal distribution applying the Shapiro-Wilk test of normality, selecting the function that gives the largest statistic (W) or p-value.

The Johnson package allows carrying out the JT according to the method described here.

4.1.4.3 Multivariate Box-Cox Transformation (MBCT)

Velilla (1993) offered a multivariate extension of the Box-Cox Transformation. Let $\lambda = [\lambda_1, \lambda_2, ..., \lambda_p]$a vector of the transformation parameters which after the following transformation $X^{(\lambda)} = \left(X_1^{(\lambda_1)}, X_2^{(\lambda_2)}, ..., X_p^{(\lambda_p)}\right)$ of the original variables, produce a multivariate normal distribution with mean ($\mu^{(\lambda)}$) and covariance ($\Sigma^{(\lambda)}$) both of the transformed variables. The λ vector is selected as the value that maximizes the log-likelihood function. See for details Velilla (1993) or Weisberg (2005).

The powerTransform function from the car package also allows computing this transformation.

Example 4.9
This example proceeds from a bivariate manufacturing process with a right-skewed distribution that can be found in rskewed data frame (Fig. 4.3).

A simple visual inspection allows verifying the presence of non-normality.

This is confirmed by the Royston (1992) and Henze and Zirkler (1990) test.

```
> HZ.test(rskewed)
p-value HZ statistic
[1] 0.00 1.95
Royston.test(rskewed)
```

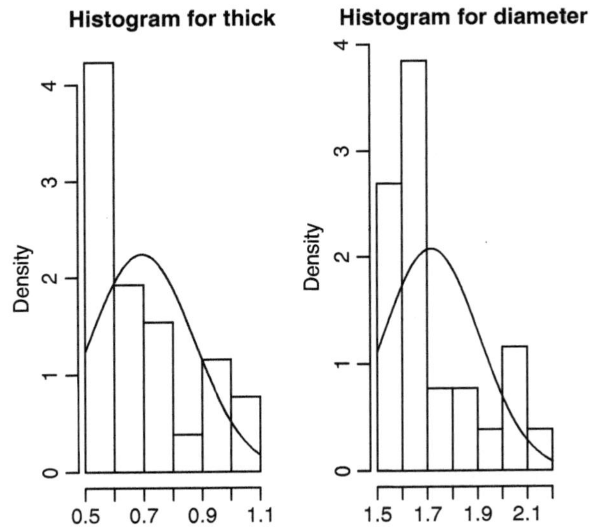

Fig. 4.3 Histogram to the rskewed data frame

test.statistic	p.value
26.75	0.00

The Mardia Test prompts the a similar result

First, applying the BCT we have:

```
> library("car")
> rskewed.bct <− matrix(0,nrow(rskewed),ncol(rskewed))
> for (i in 1 : 2){
> lambdas <− powerTransform(rskewed[,i])$lambda
> rskewed.bct[,i] <− bcPower(rskewed[,i],lambdas)}
```

Then, applying the MVN test

```
> HZ.test(rskewed.bct)
p-value HZ statistic
[1] 0.09 0.72
> Royston.test(rskewed.bct)
test.statistic p.value
6.93 0.03
```

The Royston test detects a presence of departure from normality after the transformation at $\alpha = 0.05$.

Converselly, the JT produces a success adjustment

```
> rskewed.jt <− matrix(0,nrow(rskewed),ncol(rskewed))
> for (i in 1 : 2){rskewed.jt[,i] <− RE.Johnson(rskewed[,i])$transformed}
> HZ.test(rskewed.jt)
p-value HZ statistic
[1] 0.60 0.38
> Royston.test(rskewed.jt)
test.statistic p.value
0.22 0.90
```

Finally using the MBCT

```
> rskewed.mbct <- matrix(0, nrow(rskewed), ncol(rskewed))
> lambdas <- powerTransform(rskewed)$lambda
> rskewed.mbct <- bcPower(rskewed,lambdas)
> HZ.test(rskewed.mbct)
p-value HZ statistic
[1] 0.10 0.70
> Royston.test(rskewed.mbct)
test.statistic p.value
6.81 0.03
```

This last method in the same manner to the BCT does not produce a better transformation than JT.

4.1.5 The Autocorrelation Problem

One of the requisites in control chart is the independence of the data; although, in practice this assumption is rarely checked and this could produce false alarms. It is well known that decay processes often produce variables with time dependence (see e.g.: (Mason et al. 1996) and (Mason and Young 2001) for more details.)

The presence of autocorrelation is often confirmed by plotting current observations versus preceding ones in scatter plot e.g.: x_t vs. x_{t-1}.

To illustrate this, analyze the waiting time between eruptions in the faithful dataset.

```
> f1 <- faithful[,2]
> f2 <- matrix(0, 1, length(f1))
> for (i in 1 : length(f1)){f2[i] <- f1[i + 1]}
> plot(f1, f2, xlab = "x(t)", ylab = "x(t + 1)")
```

There is strong evidence of correlation between successive pairs. The well known autocorrelation plot or correlogram introduced by Box and Jenkins (1976) is one of the most used tools to check independence (Fig. 4.4).

The autocorrelation is computed as:

$$r_h = \frac{C_h}{C_o} \tag{4.56}$$

Where

$$-1 \leq r_h \leq 1$$

$$C_h = \frac{\sum_{t=1}^{m-h} (x_t - \overline{x})(x_{t+h} - \overline{x})}{m} \tag{4.57}$$

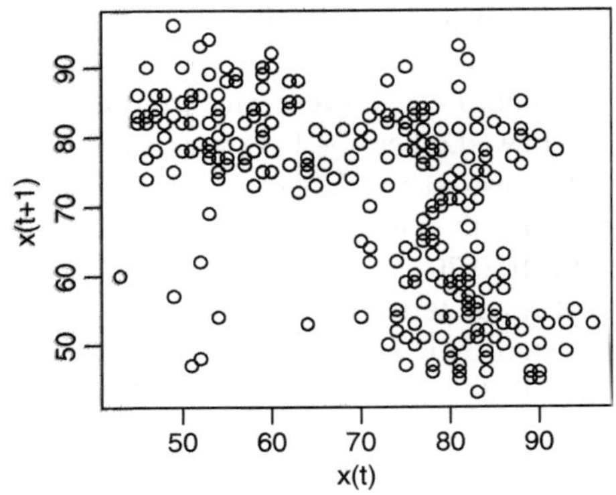

Fig. 4.4 x_t vs x_{t+1} scatter plot for the waiting time between eruptions faithful

is the covariance and

$$C_o = \frac{\sum_{t=1}^{m} (x_t - \overline{x})^2}{m} \tag{4.58}$$

the variance
and m and h the sample size and the lag respectively.

The r_h is plotted against two control limits frequently called confidence bands computed as:

$$CL = \pm \frac{Z_{1-\alpha/2}}{\sqrt{m}}. \tag{4.59}$$

When an r_h fall outside of the confidence bands, it is said that there is evidence of autocorrelation or dependence.

For more details see e.g.: Box and Jenkins (1976) or Chatfield (1989).

R provides the built-in function acf that computes the autocovariance or autocorrelation function.

Example 4.10
Coming back to the bimetal1 dataset, the marginal independence could be assessed.

```
> par(mfrow = c(3,2))
> for( i in 1 : ncol(bimetal1) ){
> par(mar = c(4.1,4.5,1,1))
> acf(bimetal1[,i],lag = 7,las = 1)}
```

Notice that when lag = 0 the correlation is 1. This can be proved easily in formula x.

There is no evidence of relation between adjacent observations; that is, there is marginal randomness.

This tool can be complemented with the use of another such as: Box-Pierce, Ljung-Box or Runs Test (Fig. 4.5).

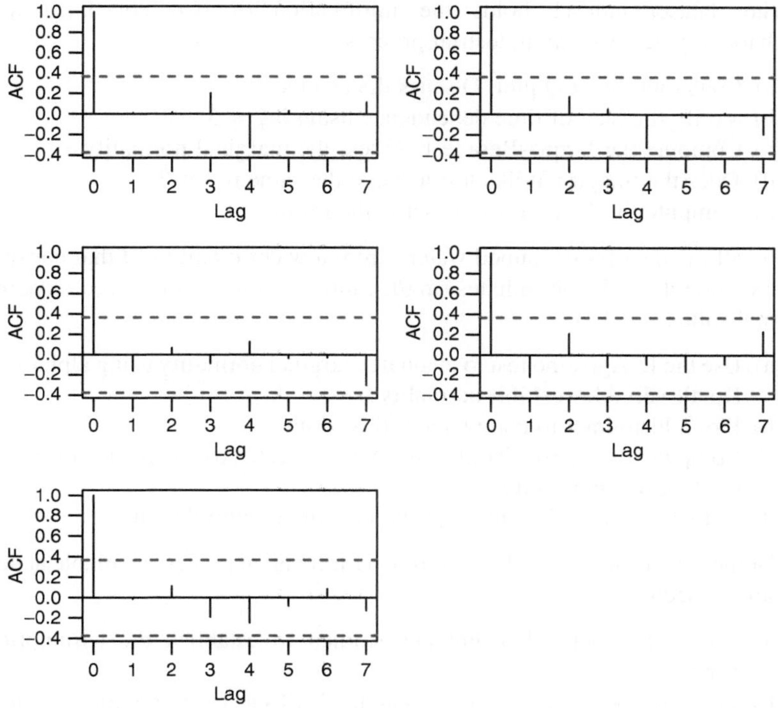

Fig. 4.5 Correlograms for each individual of the bimetal1 data frame

When time dependence is detected the problem should be addressed in two different ways: by using a specific control chart such as the proposal by Apley and Tsung (2002) and Kalagonda and Kulkarni (2004) or by modifying the data removing the autocorrelation effects. About the latter point a possible solution is to decompose it in multivariate autoregressive model and analyze the resultant residuals which should present independency and MVN (Mason and Young 2001).

4.1.6 Exercises

4.1. In Example 2.2, Sect. 2.4 a bivariate data frame called dowel1 was introduced.

(a) Perform a histogram for each quality characteristic. Does the obtained data allow foreseeing normality in data?
(b) Compute the D'Agostino test and assess the skewness, kurtosis and omnibus tests obtained.
(c) Verify the marginal normality using the Shapiro-Wilks test.
(d) Assess the multivariate normality using the Royston test.
(e) Construct the marginal ACF. Discuss the results

4.2. The dataset indust1 holds the information of two correlated quality characteristics from an industrial process.

(a) Construct the Q-Q plot. Discuss the results.
(b) Verify the lack of time dependence using lag = 5
(c) Compute the Jarque-Bera test. Assess the marginal normality.
(d) Does the Shapiro-Wilks test achieve the same results?
(e) Compute the Mardia test. Discuss the results.

4.3. Recall the data frame named water1 from a water quality test that consists on five variables (pH, phosphates (mg/L), nitrates (mg/L), dissolved oxygen and total solids (mg/L)).

(a) Use the D'Agostino test to evaluate marginal normality using alpha = 0.05. Do all variables exhibit normality?
(b) Plot a histogram to complement this result.
(c) Compute the Henze-Zirkle and Mardia test. Are there departures from multivariate normality?
(d) Construct a correlogram to prove lack of autocorrelation

4.4. For the seven variables collected from a mechanical process available at dataset named mech1:

(a) Use both graphical techniques studied to establish the assumption of normality.
(b) Compute the Jarque-Bera and Shapiro-Wilks tests and compare the results.
(c) Demonstrate the randomness using the acf function.
(d) According to the Henze-Zirkle and Royston. Assess the multivariate normality.

4.5. The gilgais dataset from MASS package presents the level of pH, electrical conductivity and chloride content from the soil in gilgai territory, New South Wales, Australia. For the first 50 samples and the characteristic pH at depth 30–40 cm and 80–90 cm:

(a) Evaluate the multivariate normality using the Henze-Zirkle and Royston tests.
(b) If any of the previous tests detects non normality presence, transform the dataset using BCT, JT and MBCT. Compare the methods according to the results obtained.
(c) Evaluate the autocorrelation level with a lag = 6 .

4.6. The Rubber data frame included in the MASS package, contain the measure from a rubber tyre accelerated testing.

(a) Test multivariate normality at alpha = 0.1.
(b) Perform a Q-Q plot. Discuss the results.
(c) Determine the presence of time dependence in all variables.

Chapter 5
Study Cases

5.1 Study Case #1. Pitching Controlling

In this study case the application of the main tools covered in this text is introduced in baseball, specifically over the pitcher performance. According to the Major League Baseball (MLB) the strike zone is "that area over home plate, the upper limit of which is a horizontal line at the midpoint between the top of the shoulders and the top of the uniform pants, and the lower level is a line at the hollow beneath the kneecap..."

It is a pentagonal prism with 20 in. (1.66 ft) of width and the height is determined by the size of the batter in the position of swinging the pitched ball. Although this height is different for each batter it normally has a dimension from 1.6 up to 3.5 ft over the home. The umpire calls strike when the pitch falls into this area and the batter does not swing.

Although the pitchers move the ball strategically in different positions of the strike zone trying the hitter not make contact with it, often the performance of the pitcher is measured by the skill to put the ball into the strike zone at high speed.

In this case we use data collected by the pitcher logs at the (MLB) database (http://gd2.mlb.com/components/game/mlb/) for the pitcher C.C. Sabathia from the New York Yankees. Two datasets were selected from games against Tampa Bay: the first on July 10, 2011 and on August 12, 2011 the second. Both are stored in the package as sabathia1 and sabathia2 respectively.

The pitcher logs provide a lot of information about each pitch but in our study we work with the start speed (given in mph) of the pitch, and the location (in feet) as it crosses the home. This last point is measured regarding a coordinate system in which the origin is at the point of the home plate. The z-axis is the vertically oriented while x-axis horizontally oriented at the catcher's right.

Only the fastball pitches are considered and each sample is a batter by averaging all the variables of pitch. Notice that a player bats several times in the play.

Performing the analysis in R.

```
> data("sabathia1")
```

E. Santos-Fernández, *Multivariate Statistical Quality Control Using R*, SpringerBriefs in Statistics 14, DOI 10.1007/978-1-4614-5453-3_5,

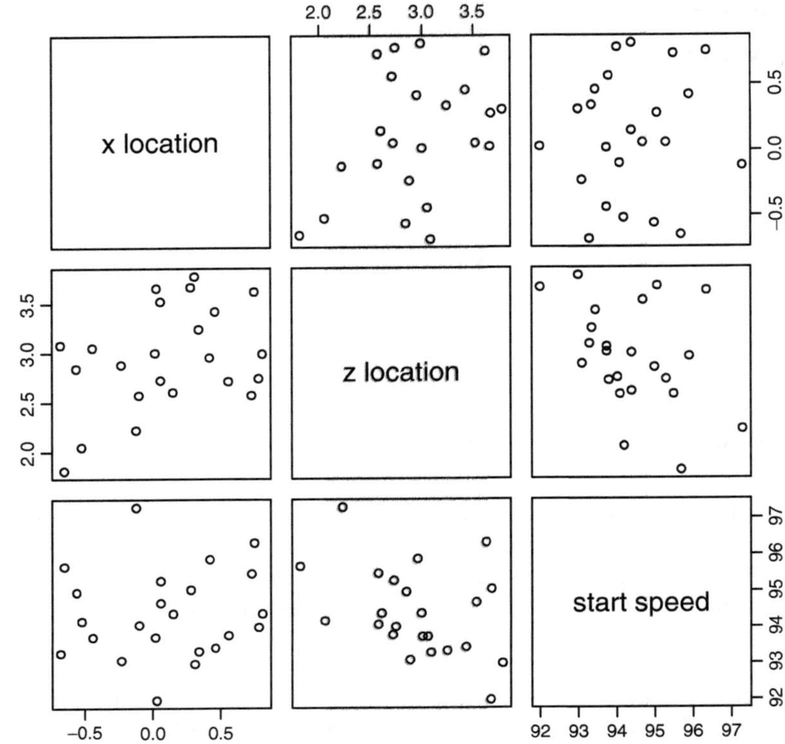

Fig. 5.1 Scatter plot matrix of *vertical* and *horizontal* location and start speed

using:

> colMeans(sabathia1); covariance(sabathia1); cor(sabathia1)

it is obtained

$$\bar{x} = \begin{bmatrix} 0.1074 \\ 2.9430 \\ 94.4108 \end{bmatrix} \quad ; \quad S = \begin{bmatrix} 0.22 & 0.09 & 0.05 \\ 0.09 & 0.27 & -0.25 \\ 0.05 & -0.25 & 1.50 \end{bmatrix} \quad \text{and} \quad r = \begin{bmatrix} 1 & 0.37 & 0.09 \\ 0.37 & 1 & -0.39 \\ 0.09 & -0.39 & 1 \end{bmatrix}.$$

Notice the direct correlation between the first two variables, being negative between the vertical position and the speed. The scatterplot matrix visually confirms this (Fig. 5.1).

> pairs(sabathia1)

An initial useful analysis can be carried out by constructing a three-dimensional scatterplot with a confidence ellipsoid (Fig. 5.2).

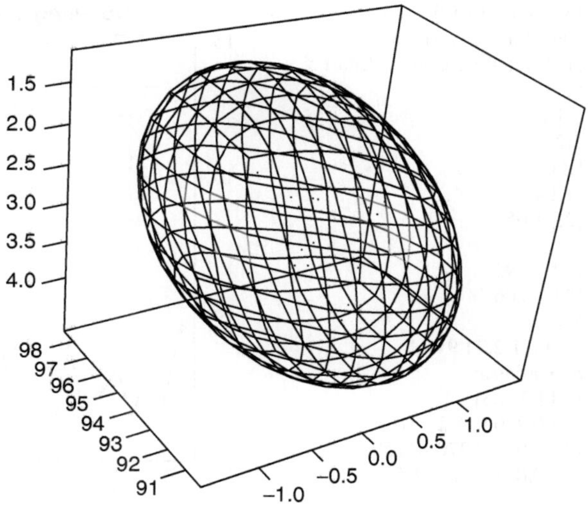

Fig. 5.2 Three dimensional scatter plot with confidence ellipsoid

```
> library(rgl)
> plot3d(ellipse3d(cov(sabathia1), centre = colMeans(sabathia1), level = 0.99),
  xlab = "", ylab = "", zlab = "",type = "wire", col = "gray1", alpha = 0.2)
> points3d(sabathia1, size = 4, cex = 2, add = TRUE)
```

By moving through the coordinates it can be seen that all observations fall inside these boundaries. No outliers are detected. Then performing a Hotelling chart (Fig. 5.3).

```
> mult.chart(type = "t2", sabathia1)
```

Since no points fall outside the UCL then there is no evidence to reject the in-control state in the process. The final score shows that.

Then, setting this first game to analyze the second game as Phase II or future observations, using the Phase I estimates of mean vector and covariance matrix as follows:

```
> colm < - nrow(sabathia1)
> vec < - (mult.chart(sabathia1,type = "t2")$Xmv)
> mat < - (mult.chart(sabathia1,type = "t2")$covariance)
Using
> data("sabathia2")
> par(mfrow = c(1,2))
> mult.chart(type = "t2", sabathia2, Xmv = vec, S = mat, colm = colm)
>
```

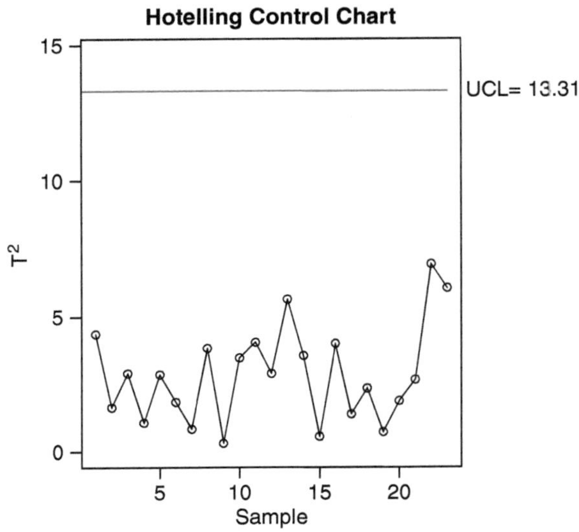

Fig. 5.3 Hotelling control chart for the sabathia1 data
[1] "Hotelling Control Chart"
$ucl
[1] 13.31
$t2
[,1]
[1,] 4.37
[2,] 1.65
...
[22,] 6.95
[23,] 6.06
$Xmv
[1] 0.11 2.94 94.41
$covariance
[,1] [,2] [,3]
[1,] 0.220 0.092 0.051
[2,] 0.092 0.270 -0.250
[3,] 0.051 -0.250 1.500

Then, R prompts:
The following(s) point(s) fall outside the control limits[1] 16 20

$'Decomposition of'	$'Decomposition of'
[1] 16	[1] 20
t2 decomp ucl p-value 1 2 3	t2 decomp ucl p-value 1 2 3
[1,] 12.5255 8.0686 0.0016 1 0 0	[1,] 0.4091 8.0686 0.5282 1 0 0
[2,] 10.7037 8.0686 0.0031 2 0 0	[2,] 9.6004 8.0686 0.0048 2 0 0
[3,] 4.2001 8.0686 0.0511 3 0 0	[3,] 0.8664 8.0686 0.3609 3 0 0
[4,] 16.8950 12.1448 0.0000 1 2 0	[4,] 13.4175 12.1448 0.0001 1 2 0
[5,] 18.1565 12.1448 0.0000 1 3 0	[5,] 1.3922 12.1448 0.2671 1 3 0
[6,] 11.3942 12.1448 0.0003 2 3 0	[6,] 15.0562 12.1448 0.0001 2 3 0
[7,] 19.4116 16.1352 0.0000 1 2 3	[7,] 22.4067 16.1352 0.0000 1 2 3

The analysis displays the points 16 and 20 beyond the UCL, i.e.: the pitcher seems to be out-of-control. The decomposition of the T^2 statistics shows how in sample 16 both locations on the horizontal and vertical axes (x) were out-of-control. In contrast, in batter number 20 only the location on the vertical causes the alarm.

In order to improve fast detection of small shifts in the process, we can compute the MEWMA and MCUSUM charts.

For instance, MEWMA detects the shifts on the mean at the 10th batter, see Fig. 5.4(b) and the MCUSUM according to (Crosier 1988) and (Pignatiello and Runger 1990) at the 9th and 10th batters respectively (Fig. 5.5).

Notice that this study does not intend to prove per se when the pitchers are in control or not. There are many other important variables to be analyzed. The aim is to propose a tool for monitoring the statistical control over the strike zone and the speed.

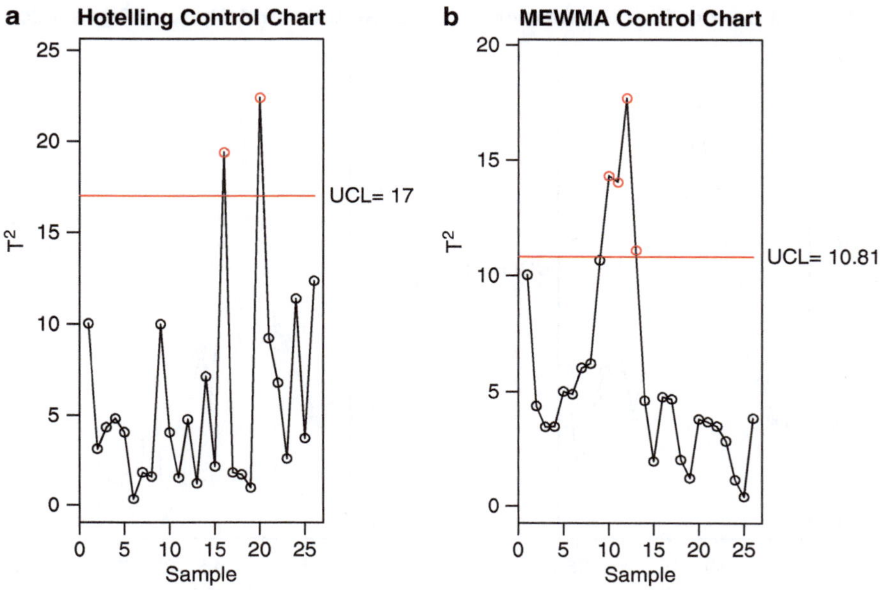

Fig. 5.4 (**a**) Hotelling and (**b**) MEWMA control chart for the sabathia2 data

Another important aspect to be considered is that although the pitcher is under statistical control over the variables measured, he could hit and the score could show a false alarm.

After that, a capability study for individual observations is performed using the umpire strike zone as specifications. The first game analyzed that was used as Phase I had in the home plate the umpire Ron Kulpa. The strike zone was constructed as the boundary rectangle of the confidence ellipse given by all the balls called strike in this game and stored in the kulpa dataset. So, using the proc.reg function the limits are computed.

```
> data("kulpa")
> LSL < - as.vector(proc.reg(kulpa, alpha = 0.1)$LPL)
> USL < - as.vector(proc.reg(kulpa, alpha = 0.1)$UPL)
```

Notice that alpha = 0.1 was used to avoid an extensive area.

```
> data("sabathia.ind")
> par(mfrow = c(1,3))
> mpci(index = "shah", sabathia.ind, LSL=LSL ,USL=USL, alpha=0.1, graph =
  TRUE)
> mpci(index = "taam", sabathia.ind, LSL = LSL ,USL = USL, alpha = 0.1, graph
  = TRUE)
> mpci(index = "pan", sabathia.ind, LSL = LSL ,USL = USL,alpha = 0.1, graph
  = TRUE)
```

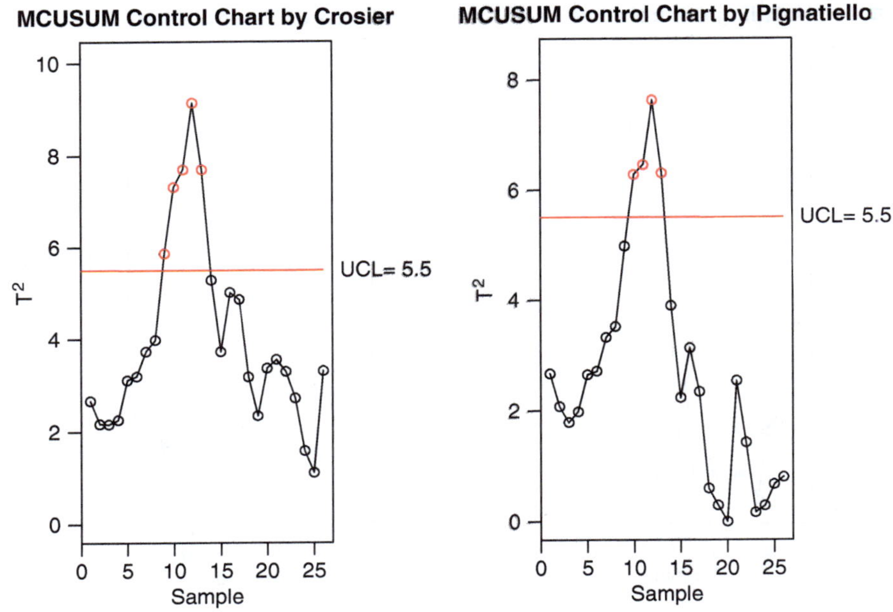

Fig. 5.5 MCUSUM control chart by (Crosier 1988) (**a**) and (Pignatiello and Runger 1990) (**b**) for the sabathia2 data

```
[1] "Shahriari et al. (1995)
    Multivariate Capability
    Vector"
$CpM
[1] 0.94
$PV
[,1]
[1,] 6.72e-05
$LI
[1] 0
```

```
[1] "Taam et al. (1993)
    Multivariate Capability
    Index (MCpm)"
$MCpm
[,1]
[1,] 0.73
```

```
[1] "Pan and Lee (2010)
    Multivariate Capability Index
    (NMCpm)"
$NMCpm
[,1]
[1,] 0.73
```

Figure 5.6 shows the output of the three indices computed. Notice the difference between the target and the process mean expressed in a extremely low value of PV in (Shahriari et al. 1995) index. The main swarm is located over the high part of the strike zone and the process region is not contained into the tolerance region, therefore LI $= 0$.

On the other hand, the area ratio of (Shahriari et al. 1995) produced a high value (0.94) while (Taam et al. 1993) and (Pan and Lee 2010) achieved lower values (0.73).

Realize that the called proportion of non conforming product in industry (in this case: balls fallen outside the umpire strike zone) is on average on one third according to MLB statistics.

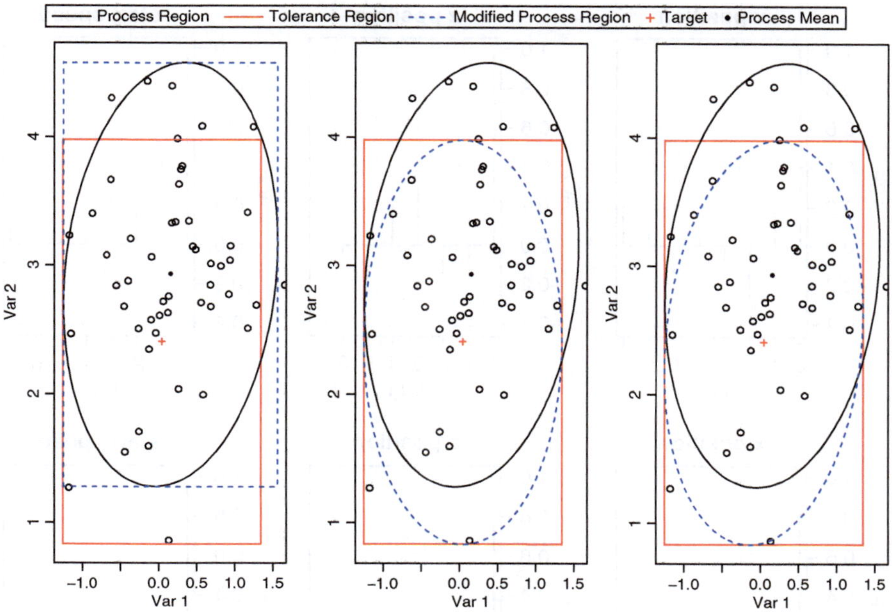

Fig. 5.6 MPCI for the sabathia1 data (Shahriari et al. 1995), (Taam et al. 1993) and (Pan and Lee 2010)

These indices could be useful to perform a comparison among pitchers and against the different umpire strike zone which varies in each game.

Finally it is checked the assumption of MVN with the Henze-Zirkler and Royston test

HZ.test(sabathia1)	HZ.test(sabathia2)
p-value HZ statistic	p-value HZ statistic
[1] 0.75 0.49	[1] 0.69 0.52
Royston.test(sabathia1)	Royston.test(sabathia2)
test.statistic p.value	test.statistic p.value
1.49 0.68	1.61 0.65

and the lack of time dependence:

```
> par(mfrow = c(2,3))
> for( i in 1 : ncol(sabathia1) ){par(mar = c(4.1,4.5,3,1))
> acf(sabathia1[,i],lag = 7,las = 1, main = colnames(sabathia1)[i])}
> for( i in 1 : ncol(sabathia2) ){ par(mar = c(4.1,4.5,3,1))
> acf(sabathia2[,i],lag = 7,las = 1, main = colnames(sabathia2)[i])}
```

Notice that no departures from normality and no autocorrelation are achieved (Fig. 5.7).

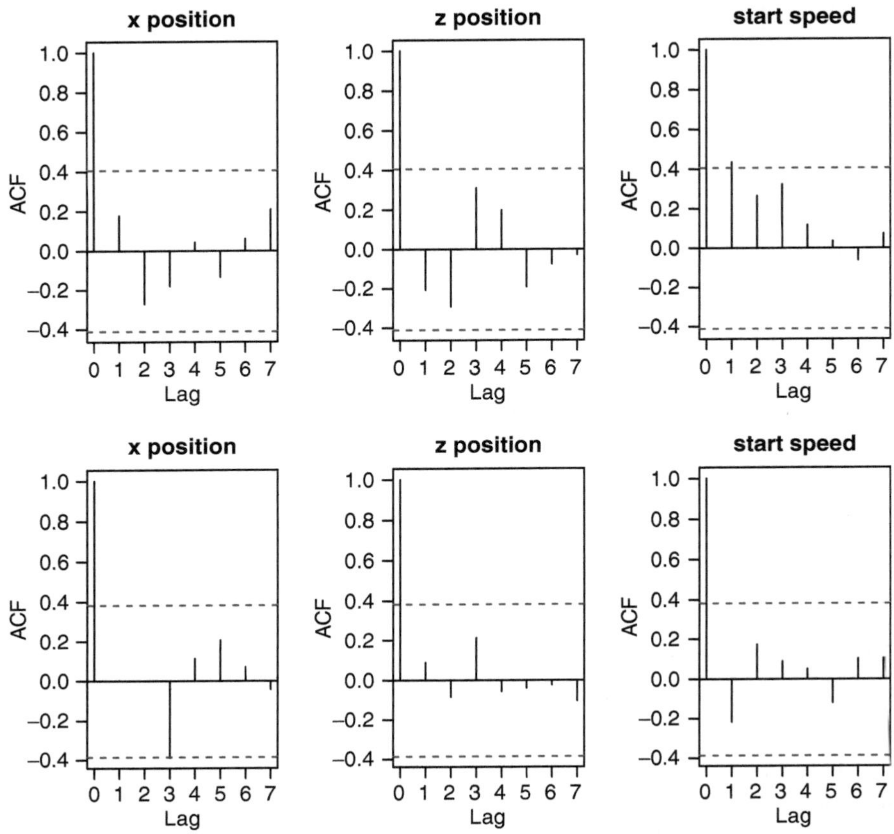

Fig. 5.7 Correlogram for both sabathia1 and sabathia2 data

This study case shows the huge spectrum of application of the multivariate quality control in which were used in combination multivariate control chart and multivariate process capability indices to evaluate the pitcher performance and the ability to fulfill the strike zone specifications.

5.2 Study Case #2. Target Archery

The target archery is a competitive sport governed by the World Archery Federation (WA) wherein the archers shoot at round target at varying distances. What is established in the Olympic Games is the 122 cm face for a distance of 70 m.

The individual competition is arranged on two stages. The first one is the ranking round in which each archer shoots 72 arrows in 12 ends of six arrows. After that,

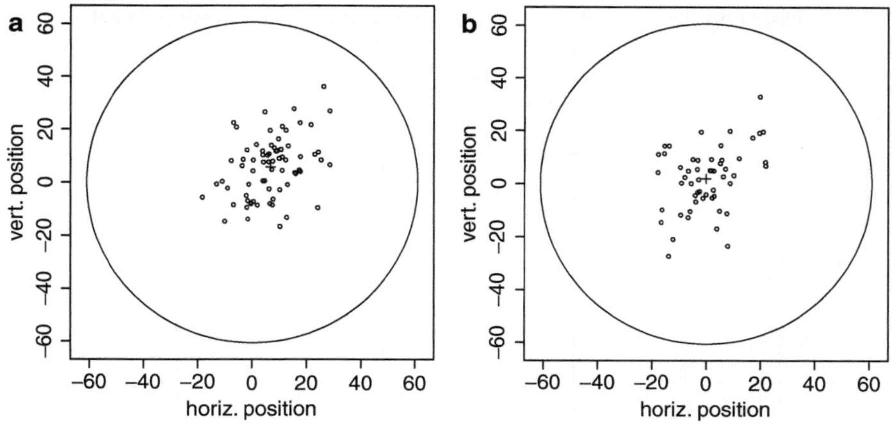

Fig. 5.8 Scatter plot with for both archery data

the second stage begins with the matches of the first ranked against the sixty-fourth, the second against the sixty-third, and so on; shooting 18 arrows in ends of three arrows. The winners move forward until completing three loops. Then the eight remaining archers continue the elimination stage shooting 12 arrows in ends of three arrows being the champion the undefeated.

The dataset called archery1 consists on the 72 shoots in ends of three arrows of the ranking round of a specific archer and the archery2, the 54 shoots of the elimination round with the same subgroup size. Notice that the information is given in x and y coordinates but in the archery competition the scoring is based on the location of the arrows over concentric rings with score values established.

The Fig. 5.8 shows the scatter plot of the individuals throws over the target of 122 cm.

```
> data("archery1")
> data("archery2")
```

The argument of the correlation function does not allows an array but using

> cor(cbind(c(archery1[,1,]),c(archery1[,2,]))) we can compute the correlation.

We have:

$$r = \begin{bmatrix} 1 & 0.37 \\ 0.37 & 1 \end{bmatrix}$$

After that the Hotelling control chart is computed for the ranking round

> mult.chart(archery1, type = "t2") then R returns:

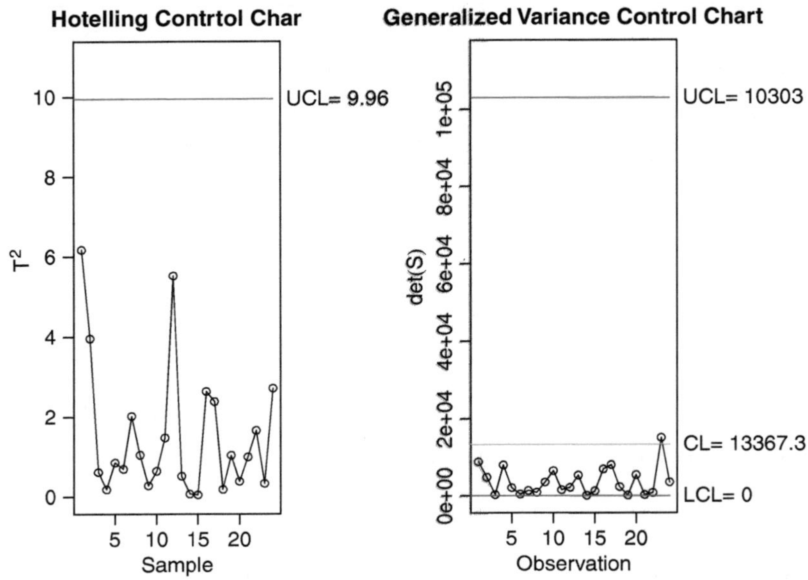

Fig. 5.9 Hotelling and generalized variance chart for archery1 data

According to the Hotelling chart the process seems to be in control since no evidence of assignable causes are presented. Now the analysis can be complemented with the generalized variance chart. This graph does not report a non-random operation either (Fig. 5.9).

```
> gen.var(archery1)
```

Suppose it is desired to use the ranking round as Phase I and to control the future observation storage on archery2 from the eliminatory (Fig. 5.10):

```
> colm < - nrow(archery1)
> vec < - (mult.chart(archery1,type = "t2")$Xmv)
> mat < - (mult.chart(archery1,type = "t2")$covariance)
> par(mfrow = c(2,2))
> mult.chart(archery2,type = "t2", Xmv = vec, S = mat, colm = colm)
> mult.chart(archery2,type = "mewma", Xmv = vec, S = mat)
> mult.chart(archery2,type = "mcusum", Xmv = vec, S = mat)
> mult.chart(archery2,type = "mcusum2", Xmv = vec, S = mat)
```

Then R prompts:

```
The following(s) point(s) fall outside the control limits[1] 18

$'Decomposition of'
[1] 18
```

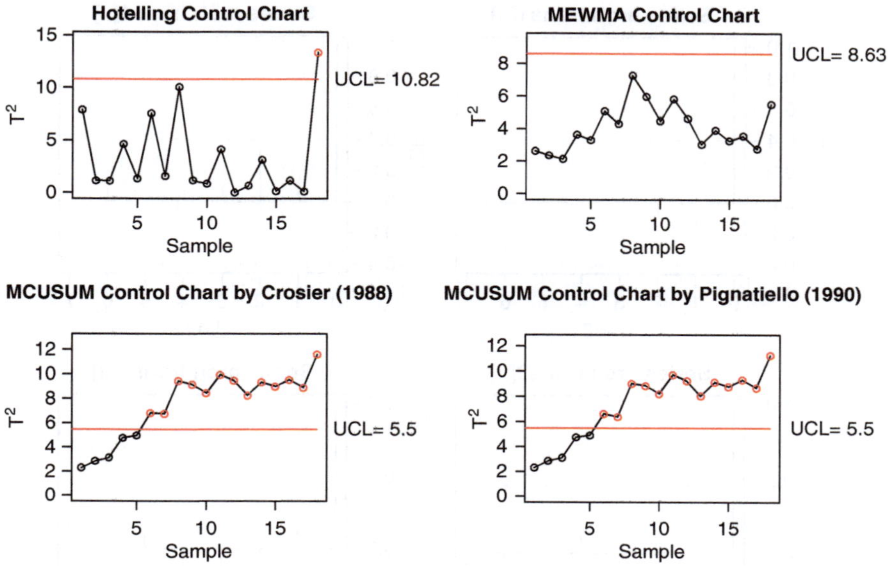

Fig. 5.10 Hotelling, MEWMA and MCUSUM control chart for archery2 data

```
t2 decomp ucl p-value 1 2
[1,] 11.4353 7.8065 0.0035 1 0
[2,] 0.0008 7.8065 0.9778 2 0
[3,] 13.3752 11.4390 0.0003 1 2
```

The Hotelling chart detects the 18th sample beyond UCL. The decomposition shows that the cause is due to a horizontal shift. While the weighted chart like the MEWMA chart does not detect non-random shifts and conversely (Crosier 1988) performs an early detection from sixth sample. The (Pignatiello and Runger 1990) chart accomplishes similar results.

To illustrate the misleading results that can be obtained with these charts when the requisites are not met and how the misuse could cause adjustment in the process when is not necessary, let us check the multivariate assumption.

```
> HZ.test(apply(archery1,1:2,mean))
p-value HZ statistic
0.07     0.73
> Royston.test(apply(archery1, 1:2, mean))
test.statistic p.value
7.02 0.03
```

```
> HZ.test(apply(archery2,1:2,mean))
p-value HZ statistic
0.43     0.40
> Royston.test(apply(archery2, 1:2, mean))
test.statistic p.value
3.49 0.18
```

As a result, the strong evidence leads to reject the multinormality in the first data. As a result a transformation is required. Using the Johnson Transformation:

```
> arch.mean1<- apply(archery1,1:2,mean); arch.mean2<- apply (archery2, 1:2,
  mean)
> arch.trans1<- matrix(0, nrow(arch.mean1), ncol(arch.mean1))
```

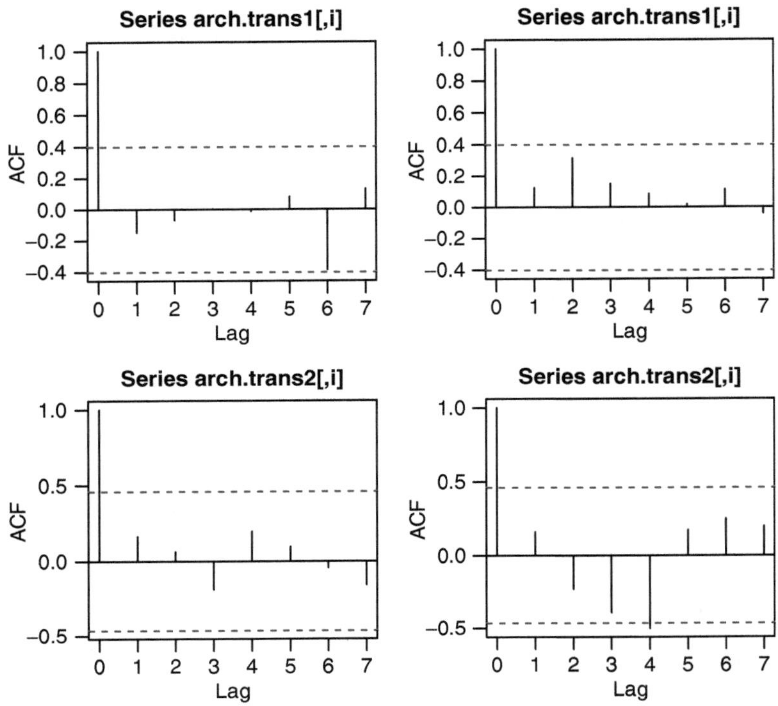

Fig. 5.11 Correlograms for both archery1 and archery2 after the transformation

```
> arch.trans2<- matrix(0, nrow(arch.mean2), ncol(arch.mean2))
> library("Johnson")
> arch.trans1[,1]<- RE.Johnson(arch.mean1[,1])$transformed; arch.trans1[,2]<-
  RE.Johnson(arch.mean1[,2])$transformed
> arch.trans2[,1]<- RE.Johnson(arch.mean2[,1])$transformed; arch.trans2[,2]<-
  RE.Johnson(arch.mean2[,2])$transformed
```

The MVN test over the transformed data is shown

> HZ.test(arch.trans1)	> HZ.test(arch.trans2)
0.32 0.48	0.99 0.15
> Royston.test(arch.trans1)	> Royston.test(arch.trans2)
test.statistic p.value	test.statistic p.value
2.48 0.29	0.44 0.80

Notice the suitable p-values achieved with this transformation. After this, the presence of autocorrelation is assessed.

```
> par(mfrow=c(2,2))
> for( i in 1 : ncol(arch.trans1) ){par(mar=c(4.1,4.5,3,1))
```

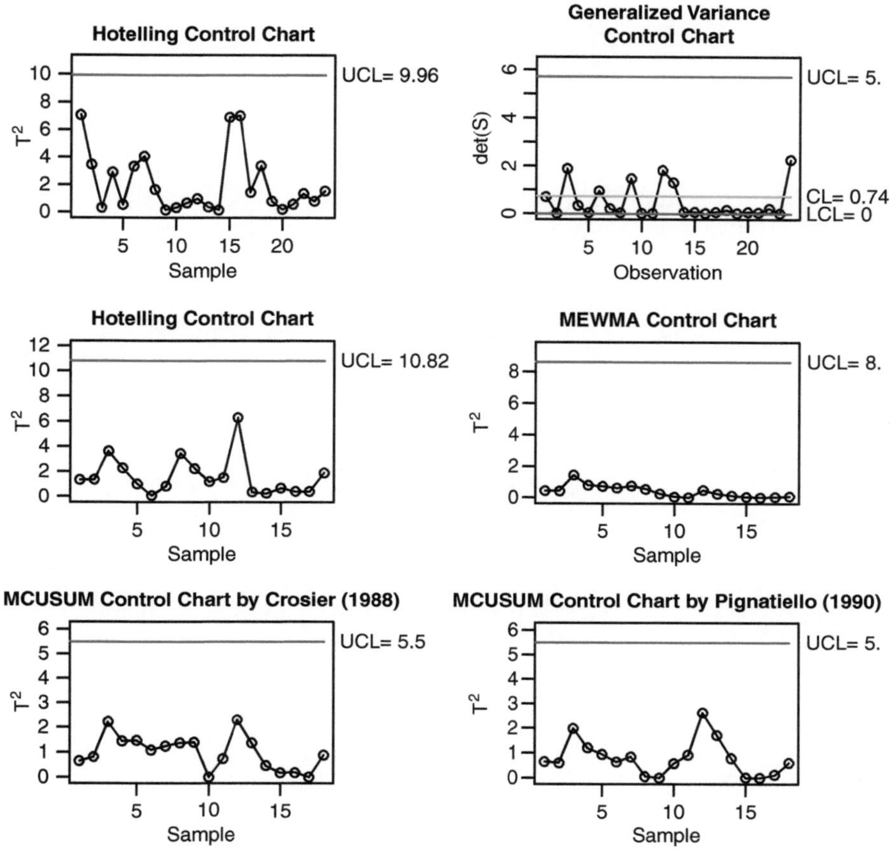

Fig. 5.12 Control charts for both archery1 and archery2

```
> acf(arch.trans1[,i],lag=7,las=1, main=colnames(arch.trans1)[i])}
> for( i in 1 : ncol(arch.trans2) ){ par(mar=c(4.1,4.5,3,1))
> acf(arch.trans2[,i],lag=7,las=1, main=colnames(arch.trans2)[i])}
```

As a result no time dependece is found. Therefore, there is no evidence to reject the randomness assumption or independence (Figs. 5.11).

Then, returning to the control chart analysis and performing the same analysis, the following results are achieved: in the ranking round the archer seems to be under statistical control since no out-of–control signal was presented. So, using this round to control the future observation (Phase II) of the elimination round, no evidence of shifts in the process was obtained. This result differs significantly to the initial analysis and shows that the non-normal presence could produce of false alarm (Fig. 5.12).

Fig. 5.12

References

Alt, F.B.: Multivariate quality control. In: Kotz, S., Johnson, N.L., Read, C.R. (eds.) Encyclopedia of Statistical Sciences, vol. 6. Wiley, New York (1985)

Anderson, T.W.: Asymptotic theory for principal component analysis. Ann. Math. Stat. **34**, 122–148 (1963)

Apley, D.W., Tsung, F.: The autoregressive T-squared chart for monitoring univariate autocorrelated processes. J. Qual. Tech. **34**, 80–96 (2002)

Bartlett, M.S.: A note on the multiplying factors for various X2 approximations. J. R. Stat. Soc. Series B. **16**, 296–298 (1954)

Bodden, K.M., Rigdon, S.E.: A program for approximating the in-control ARL for the MEWMA chart. J. Qual. Tech. **31**, 120–123 (1999)

Borror, C.M., Montgomery, D.C., Runger, G.C.: Robustness of the EWMA control chart to non-normality. J. Qual. Tech. **31**(3), 309–316 (1999)

Bothe, D.R.: A capability study for an entire product. ASQC Quality Congress Transactions. **46**, 172–178 (1992)

Box, G.E.P., Cox, D.R.: An analysis of transformations. J. R. Stat. Soc. Series B (Methodol). **26**(2), 211–252 (1964)

Box, G.E.P., Jenkins, G.: Time Series Analysis: Forecasting and Control. Holden-Day, San Francisco (1976)

Castagliola, P., Castellanos, J.-V.G.: Capability indices dedicated to the two quality characteristics case. Qual. Technol. Quant. Manag. **2**(2), 201–220 (2005)

Chan, L.K., Cheng, S.W., Spiring, F.A.: A multivariate measure of process capability. J. Model. Simulat. **1**, 1–6 (1991)

Chatfield, C.: The Analysis of Time Series: An Introduction, 4th edn. Chapman & Hall, New York (1989)

Chen, H.: A multivariate process capability index over a rectangular solid zone. Statistica Sinica. **4**, 749–758 (1994)

Chen, K.S., Pearn, W.L., Lin, P.C.: Capability measures for processes with multiple characteristics. Qual. Reliab. Eng. Int. **19**, 101–110 (2003)

Chou, Y.-M., Polansky, A.M., Mason, R.L.: Transforming non-normal data to normality in statistical process control. J. Qual. Tech. **30**(2), 133–141 (1998)

Crosier, R.B.: Multivariate generalizations of cumulative sum quality-control schemes. Technometrics. **30**(3), 291–303 (1988)

D'Agostino, R., Pearson, E.S.: Tests for departure from normality. Empirical results for the distributions of b2 and √ b1. Biometrika. **60**(3), 613–622 (1973)

D'Agostino, R.B.: Transformation to normality of the null distribution of g1. Biometrika. **57**(3), 679–681 (1970)

E. Santos-Fernández, *Multivariate Statistical Quality Control Using R*, SpringerBriefs in Statistics 14, DOI 10.1007/978-1-4614-5453-3,

D'Agostino, R.B., Belanger, A., D'Agostino Jr., R.B.: A suggestion for using powerful and informative tests of normality. The American Statistician. **44**(4), 316–321 (1990)

Doganaksoy, N., Faltin, F.W., Tucker, W.T.: Identification of out-of-control multivariate characteristic in a multivariable manufacturing environment. Comm. Stat.—Theor. Meth. **20**, 2775–2790 (1991)

Healy, J.D.: A note on multivariate CUSUM procedures. Technometrics. **29**(4), 409–412 (1987)

Henze, N., Zirkler, B.: A class of invariant consistent tests for multivariate normality. Comm. Stat.—Theor. Meth. **19**(10), 3595–3617 (1990)

Holmes, D.S., Mergen, A.E.: Improving the performance of T-square control chart. Qual. Eng. **5**(4), 619–625 (1993)

Hotelling, H.: Multivariate Quality Control. McGraw-Hill, New York (1947)

Hubele, N.F., Shahriari, H., Cheng, C.S.: A Bivariate Process Capability Vector, 299–310 (1991)

Jackson, J.E.: A User Guide to Principal Components. Wiley, New York (1991)

Jarque, C.M.: Jarque-Bera test. In: Lovric, M. (ed.) International Encyclopedia of Statistical Science. Springer, Heidelberg (2010)

Jarque, C.M., Bera, A.K.: Efficient tests for normality, homoscedasticity and serial independence of regression residuals. Econ. Lett. **6**(3), 255–259 (1980)

Jarque, C.M., Bera, A.K.: A test for normality of observations and regression residuals. Int. Stat. Rev. **55**(2), 163–172 (1987)

Johnson, N.L.: Systems of frequency curves generated by methods of translation. Biometrika. **36**, 149–176 (1949)

Juran, J.M., Godfrey, A.B.: Juran's Quality Handbook. McGraw-Hill, New York (1998)

Kalagonda, A.A., Kulkarni, S.R.: Multivariate quality control chart for autocorrelated processes. J. Appl. Stat. **31**(3), 317–327 (2004)

Kotz, S., Lovelace, C.R.: Process Capability Indices in Theory and Practice. Arnold, London (1998)

Lowry, C.A., Montgomery, D.C.: A review of multivariate control charts. IIE Trans. **27**(6), 800–810 (1995)

Lowry, C.A., Woodall, W.H., Champ, C.W., Rigdon, S.E.: A multivariate exponentially weighted moving average control chart. Technometrics. **34**(1), 46–53 (1992)

Mardia, K.V.: Measures of multivariate skewness and kurtosis. Biometrika. **57**, 519–530 (1970)

Mardia, K.V.: Applications of some measures of multivariate skewness and kurtosis for testing normality and robustness studies. Sankhya. **36**, 115–128 (1974)

Mason, R., Tracy, N., Young, J.: Monitoring a multivariate step process. J. Qual. Tech. **28**, 39–50 (1996)

Mason, R.L., Tracy, N.D., Young, J.C.: Decomposition of T-square for multivariate control chart interpretation. J. Qual. Tech. **27**, 99–108 (1995)

Mason, R.L., Young, J.C.: Multivariate Statistical Process Control with Industrial Application, 1st edn. Society for Industrial and Applied Mathematics, Philadelphia (2001)

Mecklin, C.J., Mundfrom, D.J.: An appraisal and bibliography of tests for multivariate normality. Int. Stat. Rev. **72**(1), 123–138 (2004)

MLB. http://gd2.mlb.com/components/game/mlb/.

MLB: The Strike Zone: A historical timeline http://mlb.mlb.com/mlb/official_info/umpires/strike_zone.jsp.

Montgomery, D.C.: Introduction to Statistical Quality Control, 5th edn. Wiley, New York (2004)

Murphy, B.J.: Selecting out-of-control variables with T-squared multivariate quality procedures. The Statistician. **36**, 571–583 (1987)

Nickerson, D.M.: Construction of a conservative confidence region from projections of an exact confidence region in multiple linear regression. The American Statistician. **48**(2), 120–124 (1994)

NIST / SEMATECH e-Handbook of Statistical Methods. http://www.itl.nist.gov/div898/handbook/

Page, E.S.: Cumulative sum charts. Technometrics. **3**(1), 1–9 (1961)

Pan, J.-N., Lee, C.-Y.: New capability indices for evaluating the performance of multivariate manufacturing processes. Qual. Reliab. Eng. Int. **26**(1), 3–15 (2010)
Pearn, W., Kotz, L.S., Johnson, N.L.: Distributional and inferential properties of process capability indices. J. Qual. Tech. **24**, 216–231 (1992)
Pearn, W.L., Kotz, S.: Encyclopedia and Handbook of Process Capability Indices: A Comprehensive Exposition of Quality Control Measures. World Scientific Publishing Company, Singapore (2006)
Pignatiello, J., Runger, G.: Comparisons of multivariate CUSUM charts. J. Qual. Tech. **22**(3), 173–186 (1990)
Prabhu, S.S., Runger, G.C.: Designing a multivariate EWMA control chart. J. Qual. Tech. **29**, 8–15 (1997)
Rencher, A.C.: Methods of Multivariate Analysis. Wiley, New York (2002)
Roberts, S.W.: Control chart tests based on geometric moving averages. Technometrics. **42**(1), 97–102 (1959)
Royston, J.P.: An extension of Shapiro and Wilk's W test for normality to large samples. Appl. Stat. **31**(2), 115–124 (1982)
Royston, J.P.: Some techniques for assessing multivariate normality based on the Shapiro- Wilk W. J. Roy. Stat. Soc. C (Appl. Stat). **32**(2), 121–133 (1983)
Royston, J.P.: Approximating the Shapiro-Wilk W-Test for non-normality. Stat. Comput. **2**(3), 117–119 (1992)
Royston, J.P.: Remark AS R94: a remark on Algorithm AS 181: the W test for normality. J. Roy. Stat. Soc. C (Appl. Stat). **44**(4), 547–551 (1995)
Santos-Fernández, E., Scagliarini, M.: MPCI: an R package for computing multivariate process capability indices. J. Stat. Softw. **47**(7), 1–15 (2012)
Scagliarini, M.: Multivariate process capability using principal component analysis in the presence of measurement errors. AStA Advances in Statistical Analysis. **95**(2), 757–765 (2011)
Shahriari, H., Hubele, N.F., Lawrence, F.P.: A multivariate process capability vector. In: Proceedings of the 4th Industrial Engineering Research Conference, vol. 1, pp. 304–309. (1995)
Shapiro, S., Wilk, M.: An analysis of variance test for normality. Biometrika. **52**, 591–611 (1965)
Shinde, R.L., Khadse, K.G.: Multivariate process capability using principal component analysis. Qual. Reliab. Eng. Int. **25**(1), 69–77 (2008)
Slifker, J.F., Shapiro, S.S.: The Johnson system: selection and parameter estimation. Technometrics. **22**(2), 239–246 (1980)
Sullivan, J.H., Woodall, W.H.: A comparison of multivariate control charts for individual observations. J. Qual. Tech. **28**, 398–408 (1996)
Taam, W., Subbaiah, P., Liddy, W.J.: A note on multivariate capability indices. J. Appl. Stat. **20**, 339–351 (1993)
Tano, I., Vännman, K.: Comparing confidence intervals for multivariate process capability indices. Qual. Reliab. Eng. Int. **28**(4), 481–495 (2011)
Testik, M.C., Runger, G.C.: Mining manufacturing quality data. In: Ye, N. (ed.) Handbook of Data Mining. Lawrence Erlbaum Associates Publishers, New Jersey (2003)
Thode, H.C.: Testing for Normality. Marcel Dekker, New York (2002)
Thode, H.C.: Normality tests. In: Lovric, M. (ed.) International Encyclopedia of Statistical Science. Springer, New York (2010)
Tracy, N.D., Young, J.C., Mason, R.L.: Multivariate control charts for individual observations. J. Qual. Tech. **24**, 88–95 (1992)
Velilla, S.: A note on the multivariate Box-Cox transformation to normality. Stat. Probab. Lett. **17**, 259–263 (1993)
Venables, W.N., Ripley, B.D.: Modern Applied Statistics with S. Fourth Edition, 4th edn. Springer, New York (2002)
Wang, C.H.: Constructing multivariate process capability indices for short-run production. Int. J. Adv. Manuf. Technol. **26**, 1306–1311 (2005)

Wang, F.K., Chen, J.C.: Capability index using principal components analysis. Qual. Eng. **11**, 21–27 (1998)

Wang, F.K., Hubele, N., Lawrence, F.P., Miskulin, J.D., Shahriari, H.: Comparison of three multivariate process capability indices. J. Qual. Tech. **32**, 263–275 (2000)

Weisberg, S.: Applied Linear Regression, 3rd edn. Wiley/Interscience, New York (2005)

Wierda, S.J.: A multivariate process capability index. In: ASQC Quality Congress Transactions, pp. 342–348. (1993)

Wierda, S.J.: Multivariate statistical process control—recent results and directions for future research. Statistica Neerlandica. **48**, 147–168 (1994)

Woodall, W.H., Ncube, M.M.: Multivariate CUSUM quality-control procedures. Technometrics. **3**(3), 285–292 (1985)

Woodall, W.H., Controversies and contradictions in statistical process control. Journal of Quality Technology **32**(4), 341–350 (2000)

Xekalaki, E., Perakis, M.: The use of principal component analysis in the assessment of process capability indices. In: Proceedings of the Joint Statistical Meetings of the American Statistical Association, The Institute of Mathematical Statistics, The Canadian Statistical Society. New York. Marcel Dekker, New York (2002)

Yum, B.-J., Kim, K.-W.: A bibliography of the literature on process capability indices: 2000–2009. Qual. Reliab. Eng. Int. **27**(3), 251–268 (2012)

Index

E. Santos-Fernández, *Multivariate Statistical Quality Control Using R*,
SpringerBriefs in Statistics 14, DOI 10.1007/978-1-4614-5453-3,

MIX
Papier aus verantwortungsvollen Quellen
Paper from responsible sources
FSC® C105338

If you have any concerns about our products,
you can contact us on
ProductSafety@springernature.com

In case Publisher is established outside the EU,
the EU authorized representative is:
Springer Nature Customer Service Center GmbH
Europaplatz 3, 69115 Heidelberg, Germany

Printed by Libri Plureos GmbH
in Hamburg, Germany